13th International Symposium on Modern Optics and Its Applications (ISMOA 2021)

Journal of Physics: Conference Series Volume 2274

Online
1 – 4 August 2021

ISBN: 978-1-7138-5526-2
ISSN: 1742-6588

Printed from e-media with permission by:

Curran Associates, Inc.
57 Morehouse Lane
Red Hook, NY 12571

Some format issues inherent in the e-media version may also appear in this print version.

Institute of Physics
Dirac House, Temple Back
Bristol BS1 6BE UK

Phone: 44 1 17 929 7481
Fax: 44 1 17 920 0979

techtracking@iop.org

Additional copies of this publication are available from:

Curran Associates, Inc.
57 Morehouse Lane
Red Hook, NY 12571 USA
Phone: 845-758-0400
Fax: 845-758-2633
Email: curran@proceedings.com
Web: www.proceedings.com

TABLE OF CONTENTS

PREFACE

Since 2020, we have had to adapt our activities due to the CoVid-19 pandemic. The prolonged restriction of movement has hampered our scientific research activities. We long for a venue where we can share and discuss optics and photonics, thus we are glad that the Indonesian Optical Society (InOS) biannual main meeting, the 13th International Symposium on Modern Optics and Its Applications (ISMOA) had been successfully held on 2-4 August 2021 virtually through video conference. With a theme of "Photonics for Better Lives", this symposium was organized in collaboration with the Department of Electrical Engineering, Universitas Pelita Harapan, together with Institut Teknologi Bandung and Institut Teknologi Sepuluh Nopember as one of the International Day of Light (IDL) 2021 activities in Indonesia.

The ISMOA meeting series is the flagship conference of the Indonesian Optical Society (InOS) which has been successfully organized twelve times. Since the first ISMOA held in 2001, for ten times, these meeting series have always been held at Institut Teknologi Bandung (ITB) in Bandung, Indonesia, wherein 2011 the Indonesian Optical Society was founded. Thereafter, to fulfil the mission of the society to disseminate the science of optics and its important contributions to the advances of science and technology in general, a rotating scheme of hosting the meeting by other universities and cities in Indonesia was started. This 13th meeting is the third time that the meeting was held outside of ITB. This 13th symposium was highlighted by 13 distinguished international scientists who delivered invited presentations on their latest research, namely

Aslam BAIG (National Centre for Physics, Pakistan)
Humeyra CAGLAYAN (Tampere Univ., Finland)
Martijn DE STERKE (Univ. Sydney, Australia)
Suganda JUTAMULIA (Omnivision Technologies, Inc., USA)
Takashi KURODA (National Institute for Material Science, Japan)
Laura M. LECHUGA (Catalan Institute of Nanoscience and Nanotechnology, Spain)
David MARPAUNG (Univ. Twente, The Netherlands)
Thomas PERTSCH (Friedrich-Schiller University Jena, Germany)
Jonghwa SHIN (Korea Advanced Institute of Science and Technology, South Korea)
Kaoru TAMADA (Kyushu University, Japan)
Nelson TANSU (Univ. Adelaide, Australia)
Thomas WHITE (Australian National University, Australia)
Yuanlin ZHENG (Shanghai Jiao Tong University, P. R. China)

out of which, 5 are Fellows of Optica.

Optics and Photonics have been important in many different areas, in particular recently in the current battle with the CoVid pandemic. In this context, the symposium was highlighted by the talk of Prof. Laura Lechuga on Nanophotonic Biosensor for CoVid-19 Detection.

This 13th meeting also marks the inauguration of the May-On Tjia Lecture to commemorate the late Prof. May-On Tjia's lifelong contribution to physics and optics in Indonesia in particular. The first May-On Tjia Lecturer was Prof. David Marpaung who introduces the new optical device termed Programmable Photonic Chip.

As is the norm of the ISMOA meetings, these series of symposia are always preceded by tutorial sessions that are aimed to equip the participants with some recent information on new concepts and technologies in optics and photonics that may be useful to young scientists and students who want to start their endeavour in research. The tutorial sessions feature 4 lectures delivered by experts on various topics of interest to the ISMOA community.

There were 35 contributed presentations where 15 were delivered as oral presentations and the rest in the form of poster presentations due to limited available time. Each oral presentation was given 15 minutes of talk including Q & A, while each poster presentation was given 3 minutes preview in a special plenary session in addition to a multimedia presentation on youtube and in virtual breakout rooms with the opportunity of interactions between participants and presenters. The symposium was attended by a total of 84 speakers and attendees coming from 13 different countries (Indonesia, Australia, China, Finland, Germany, Japan, Malaysia, Netherlands, Pakistan, South Korea, Spain, UK, and the United States).

Out of the presented papers, the Review Committee has selected 3 papers for further submission to the Journal of Nonlinear Optical Physics and Materials and 13 papers that pass the review process to be published in The Journal of Physics: Conference Series. We thank The International Scientific Committee, Symposium Chairs, Reviewer Board, and Local Organizing Committee for their professional work guaranteeing high quality of the symposium.

International Scientific Program Committee

Symposium Chairs

Reviewer Board

We gratefully acknowledge the financial support from our sponsors that enable us to successfully execute the symposium. We also appreciate the endorsement from Optica (formerly OSA).

Chair of the Organizing Committee

Dr. Henri P. Uranus

Chair of the Scientific Program Committee

Assoc. Prof. Alexander A.P. Iskandar

Journal of Physics: Conference Series **2274** (2022) 011001 doi:10.1088/1742-6596/2274/1/011001

May-On Tjia Lecture

Prof. May-On Tjia, the initiator of the ISMOA meeting series and co-founder of the Indonesian Optical Society, passed away on 5 November 2019. The late Prof. May-On Tjia received his Bachelor of Physics from Institut Teknologi Bandung (ITB), Bandung, Indonesia in 1962, and continued at Northwestern University in the United States of America, where he obtained his Ph.D. in 1969 in particle physics. Gradually, his research interests shifted to conjugated polymers, superconductors, laser spectroscopy, and modern optics. He worked on these topics productively until the end of his life. Prof. May-On Tjia never even considered stopping doing science: after his retirement as a Professor from ITB in 2005, he still informally supervised students in the Physics of Magnetism and Photonics Research Divisions and was involved in research on laser-induced breakdown spectroscopy in the Maju Makmur Mandiri Research Center, Jakarta.

Prof. May-On Tjia was a Senior Member of the Optica (formerly OSA) and received several national awards. In 2007, he was the recipient of the "Anugrah Sewaka Winayaraha" award from the Indonesian government for distinguished service to higher education in Indonesia. In 2009 he was awarded the "Ganesa Cendikia Widya Adiutama" award from the Institut Teknologi Bandung for unusually wide-ranging internationally published research contributions in material physics. He was the recipient of "Achmad Bakrie Award 2012" award from the Freedom Institute for exceptional contribution to the scientific research output of Indonesia and the development of research culture in Indonesia, and of the "Sarwono Award 2016" award from the Indonesian Institute of Sciences (LIPI) for outstanding contribution to the results of scientific research and dedication in developing a research culture in Indonesia.

The May-On Tjia Lecture series is created to commemorate his scientific legacy which is based on more than 50 years (1966-2019) of research. During times when only scant research funding was available, Prof. Tjia did not keep his research brain idle (this was the term he used to use), and never ceased to do research, so he was ready whenever opportunities arose. The first May-On Tjia Lecture on Modern Optics was organized by the Physics of Magnetism and Photonics Research Division of the Institut Teknologi Bandung in collaboration with the Organizers of the 13th International Symposium on Modern Optics and Its Applications 2021. The May-On Tjia Lecture on Modern Optics will be incorporated in the program of this symposium and in future ISMOA symposia.

Considering the broad spectrum of Prof. Tjia's interests that encompass Optics and Magnetics, the May-On Tjia Lecture will be delivered by scientists working in one of these areas, and who share Prof. Tjia's values of building a deep-rooted research culture in his/her environment and demanding high-quality research from themselves and from their collaborators.

The 2021 May-On Tjia Lecturer was **Prof. David Marpaung** of the University of Twente, the Netherlands, talking on "Programmable Photonic Chips"

Peer Review Statement

All papers published in this volume have been reviewed through processes administered by the Editors. Reviews were conducted by expert referees to the professional and scientific standards expected of a proceedings journal published by IOP Publishing.

- **Type of peer review:** Single Anonymous
- **Conference submission management system:** Morressier
- **Number of submissions received:** 17
- **Number of submissions sent for review:** 17
- **Number of submissions accepted:** 13
- **Acceptance Rate (Submissions Accepted / Submissions Received × 100):** 76.5
- **Average number of reviews per paper:** 2
- **Total number of reviewers involved:** 8
- **Contact person for queries:**
  Name: Dr. Alexander A. Iskandar
  Email: iskandar@fi.itb.ac.id
  Affiliation: Institut Teknologi Bandung, Indonesia

Green synthesis of silver nanoparticles from anthocyanin extracts of purple cabbage (*brassica oleracea var capitata*) and its characteristics for dye-sensitized solar cells (DSSC) application

Waode Sukmawati Arsyad[1,*], Febryanti Cassandra[1], Muh. Asharuddin[1], Suryani Suere[2], La Ode Ahmad Nur Ramadhan[3] and Rahmat Hidayat[4]

[1] Physics Department, Faculty of Mathematics and Natural Sciences, Halu Oleo University, Indonesia
[2] Faculty of Pharmacy, Halu Oleo University, Indonesia
[3] Chemistry Department, Faculty of Mathematics and Natural Sciences, Halu Oleo University, Indonesia.
[4] Physics Department, Institut Teknologi Bandung, Indonesia.

[*]Corresponding author: wdsukmawati@uho.ac.id

Abstract. We extracted natural dye from purple cabbage (PCE) and green-synthesized the silver nanoparticle (AgNPs) from this extract for the light-harvesting material in DSSC. The anthocyanin was extracted with the maceration method for 24 hours with solvent variation (distilled water (DW), ethanol and methanol), followed by synthesized AgNPs. From FTIR, we found that the absorption area of anthocyanin at 1629 cm^{-1} correspond with stretch vibration of C=O shifted to 1608 cm^{-1} in AgNP, indicates the presence of amine group or flavanones group. The PCEs showed an amorphous non-Bragg diffraction peak at a range of (15-25)°. AgNP's diffractogram showed an intense peak at around 37.94° originates from Ag. Visible light range absorption observed, peaked at (421, 480, 550, and 966) nm, (415 and 544) nm, and (409 and 544) for PCEs in DW, ethanol, and methanol, respectively. The peak at 550 nm that comes from aglycone group in anthocyanin would be disappeared in the AgNPs that was made by dissolving $AgNO_3$ in DW, ethanol, and methanol. The energy gap was (2.37, 3.00, and 3.03) eV for PCEs in DW, ethanol, and methanol, respectively, and (3.20, 3.30, and 3.31) eV for AgNPs. The reduction current originated from the Faradaic process (dark and irradiated condition) was detected in the extract with DW and ethanol as solvent, while it was not noticeable in methanol. This reduction current was detected for all solvents in AgNP positioned at a positive voltage of 0.2V. To be used as a light absorber in DSSC, one must choose the dye solution with the smallest energy gap without or small reduction current.

1. Introduction

Nanoparticles (NPs) which refers to 1 – 100 nm [1] sized particles is emerging materials that offer a larger surface area to volume ratio [2] and has a wide range of applications starts from antimicrobials [3, 4, 5, 6], drugs [2], biosensors [7] such us HIV early detection [8], contaminant removal [9], optoelectronic devices [10], and energy storage devices [11, 12]. Based on the morphology, size, and chemical properties, NPs can be divided into categories, such as carbon-based NPs, metal NPs, ceramic NPs, semiconductor NPs, polymeric NPs, and liquid-based NPs [13]. Among others, metal NPs show remarkable optical properties due to the localized surface plasmon resonance (LSPR) characteristics [13]. In general, there are three requirements to synthesize NPs, specifically selection of the proper solvent, reducing agent, and a harmless material for stabilization [14]. Physical and chemical methods have already used for obtaining more pure metal NPs but, it is destructive, expensive, and it is not environmental friendly. To overcome the problem related to the toxicity from conventionally producing NPs, biological methods or green synthesis methods provides eco-friendly routes for synthesizing metal NPs, using microorganism such as bacteria [15] and fungi [16], enzymes [17], algae and plants [2, 13].

Plants have been used for various NPs syntheses since their extract contains phytochemicals (polyols, terpenoids, and polyphenols) [18] that can act as both stabilization and reducing agent [14, 19]. Plant utilization in NPs synthesis is advantageous compared to other biological sources. For example, the utilization of microbes will consume more time regarding the culturing process, the difficulties in maintaining cell structures, including size distribution, shape, and crystallinity [20]. Moreover, the use of plants opens up the possibility for a large-scale NPs synthesis [2, 21], since various parts of plants can be used such as leaves [22], flowers [23], fruits [24], roots [3], stems [25], and seeds [26]. Using this green concept, several NPs have been synthesized, including gold (Au), silver (Ag), zinc oxide (ZnO), and iron (Fe) [14] using plant extracts such as orange [6], lemons [27], pepper, Aloe vera, Nigella sativa, Pulicaria glutinosa, Justicia glauca, Mimusops elengi L, coffee, Lawsonia inermis [28], and purple or red cabbage [29, 30].

Red or purple cabbage extract has been used as the dye food coloring [31] and also as the light harvested material in DSSCs [32, 33, 34, 35]. Furthermore, green-synthesized AgNPs from red or purple cabbage also has been reported. However, the application is limited as an antioxidant [29] and antimicrobial [30]. Herein, we analyzed the chemical and physical properties of the red or purple cabbage extract and the resulting silver nanoparticles (AgNPs), associated with the possibility of AgNPs to be used as the light harvested materials in DSSCs. We synthesized AgNPs using red or purple cabbage extract, where the latter will act as both reducing and stabilizing agents, which commonly known as the green synthesis or biosynthesis method, since no need for any chemical reducing agent. The reaction process was carried out in three different solvents, specifically ethanol, methanol, and distilled water.

2. Experimental

2.1. Crude anthocyanin's extraction from purple cabbage

Fresh purple cabbage (PC) tore into sheets, washed with water, and dried in the air at room temperature. The PC was cut into small pieces and crushed using a Philip blender. 20 gr of PC was added to 20 ml of distilled water in a beaker and stirred using a magnetic stirrer for 10 minutes. The PC solution was then kept for 24 hours at room temperature (maceration process) on a closed beaker and then filtered using Whattman paper to remove the residual part. We repeated the same procedures for ethanol and methanol solvent. We kept the PCE's solution in a refrigerator before being used.

2.2. Synthesize of silver nanoparticles (AgNPs)

AgNPs synthesize process following the procedures reported in [30] with a modification. $AgNO_3$ with 1 mM concentration was dissolved in 100 ml of distilled water and stirred. 10 ml of purple cabbage extract (PCE) solution in distilled water then added into 90 ml of $AgNO_3$ solution, followed by incubation process in a closed beaker for 15 minutes, and the residual part filtrated using Whatman paper. We repeated the same procedures for ethanol and methanol. We kept the AgNPs solution in a refrigerator before being used.

2.3. TiO_2 electrode preparation for linear sweep voltammetry (LSV) measurement

We used cleaned FTO (*Fluorine Tin Oxide*) conductive glass (R = 15 Ω/cm^2 from Solaronix) as substrate. Four TiO_2 transparent layers (Ti-Nanooxide T/SP from Solaronix) and one TiO_2 reflector layer (18-NRT from *Dye Sol*) were deposited on top of FTO by a screen printing method. Each layer was heated on a hot plate at 80°C for 5 minutes. The second layer was coated on top of it and so on, until the last (fourth layer). At the latter, the temperature gradually increased up to 500°C and was held for 30 minutes at this temperature. We deposited the TiO_2 scattering layer on top of the TiO_2 transparent, followed by an annealing process at the hot plate. The temperature also gradually raised up to 500°C. The resulting TiO_2 electrode was immersed in the PCE and AgNPs solutions for 24 hours.

2.4. Characterization
PCE and AgNPs absorption spectrum was measured with UV-Vis spectrometer OCEAN OPTICS 2000, while their functional group, crystal structure, and photo-catalysis properties were characterized using Shimadzu Fourier transform infrared (FTIR), XRD, and LSV, respectively. The existence of the AgNPs was confirmed using JEOL JEM 1400 Transmission electron microscopy (TEM).

For LSV measurement, 0.1M $NaNO_3$ solution was used as the electrolyte, with TiO_2 immersed in PCE and AgNPs solution as a working electrode and platinum wire as the reference electrode. The voltage was from -1.0 V to 1.0 V.

3. Results and discussions

3.1. PCE and AgNPs color characteristics
The color of PCE and AgNP in all solvents are shown in Figure 1(a, b, c, and d). A_0, E_0, and M_0 refer to the PCE solution in distilled water, ethanol, and methanol, respectively. A_1, A_2, E_1, and M_1 refer to the AgNPs solution where the silver nitrate dissolved in distilled water before mixed with A_0, E_0, and M_0, respectively. Meanwhile, E_2 refers to the AgNPs solution where the silver nitrate dissolved in ethanol before mixed with E_0. M_2 refers to the AgNPs solution where the silver nitrate dissolved in methanol before mixed with M_0.

(a) (b) (c) (d)

Figure 1. The color of PCE in all solvents (a), the color comparison of PCE and AgNPs solution in distilled water (b), ethanol (c), and methanol (d).

From Figure 1(a), the PCE solution from all solvents shows a similar color, reddish-purple, but with a different color density. The color produced from PCE extracted with distilled water is more concentrated than the color produced in ethanol and methanol. Figures 1(b), 1(c), and 1(d) show the color evolution from PCE to AgNP. The color change was observed when we use different solvents to dissolve silver nitrate. When we use distilled water to dissolve silver nitrate, the resulting AgNPs solution was light pink-purple (A_1) and opaque pink (A_2). Meanwhile, the color changed to light purple in E_1 and M_1. When we used ethanol and methanol to dissolve silver nitrate, the color of AgNPs solution change to a very faded-pink (E_2) and clear (M_2), respectively.

In terms of color stability, all PCE shows no change in color, and no precipitation was observed after five days at room temperature (Figure 2a). Meanwhile, AgNPs solution A_1 and A_2 show very distinguished colors from light pink-purple to light brown (A_1** and A_2** in Figure 2b). The AgNPs solution E_1 and M_1 show a slight change in color from light purple to dark purple (E_1** and M_1** in Figure 2c), and AgNPs solution E_2 and M_2 show no color change (Figure 2d). For all AgNPs samples, we observed the precipitate in the solution. From these result, one can say that AgNPs that was synthesized using ethanol and methanol shows more stable color.

(a) (b) (c) (d)

Figure 2. The color of PCE in all solvents (a), the color comparison of PCE and AgNPs solution in distilled water (b), ethanol (c), and methanol (d). The pictures were captured after the samples left for five days at room temperature.

3.2. Functional group characterization of PCE and AgNPs

FTIR spectroscopy measurement was conducted to observe the change of chemical bonds in the PCE and AgNP. The FTIR spectra of PCE and AgNP in ethanol solution (E_0 and E_2) were shown in Figure 3.

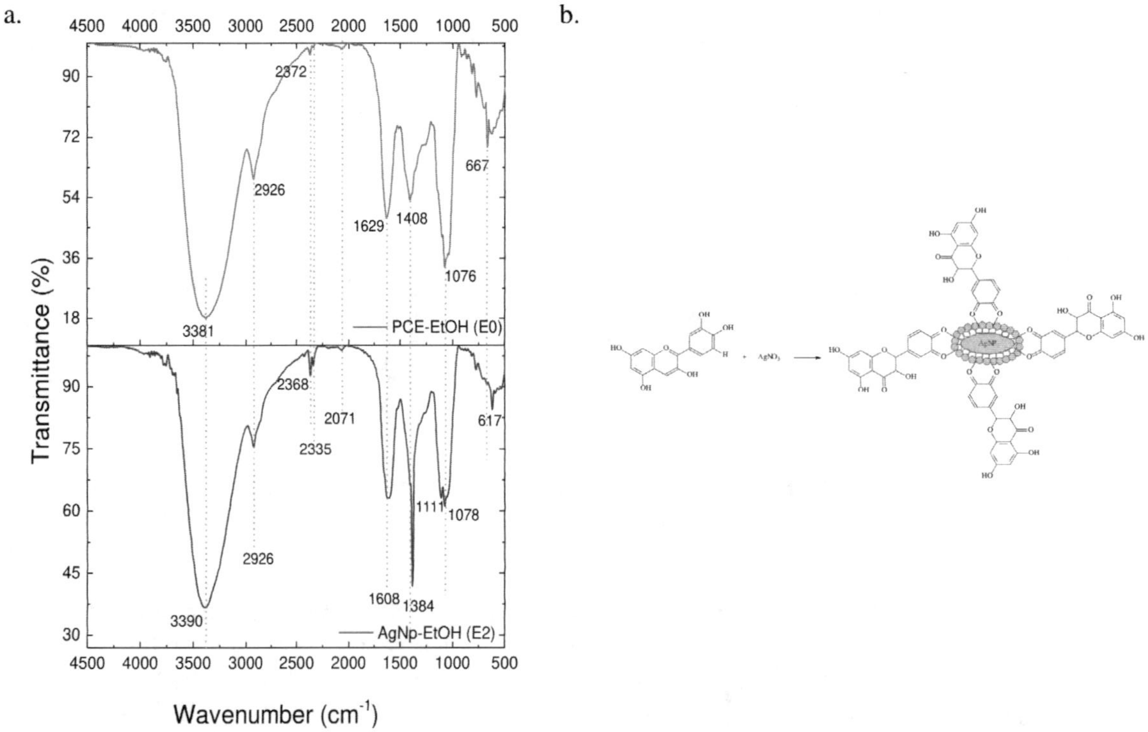

Figure 3. (a) The FTIR spectra of PCE (red lines) and AgNP (blue lines) in ethanol solution (E_0 and E_2), and (b) the proposed reaction between cyanidin molecules in anthocyanin with silver nitrate to form silver nanoparticles (AgNP)

PCE's FTIR spectra shown a sharp absorption peak at 3000-3500 cm^{-1}, 2926 cm^{-1}, and 1629 cm^{-1} corresponds to O-H bond, C-H_3 bond (methyl group), and C=O stretching vibration, respectively. The absorption at 1629 cm^{-1} might also arise from stretch vibration of carboxyl ions (COO-) (symmetry or asymmetrical), indicates the presence of carboxylate acid group, ester or carbonyl group [36]. The high and low intensity of bend vibration from C-O-C observed at a frequency of 1076 cm^{-1} and 1261 cm^{-1}, respectively. It was an indication of the presence of the carbohydrate group [37, 38].

The FTIR spectra of AgNPs synthesized from PCE in ethanol shows absorption peak at 1078 cm^{-1} originated from C-O stretch alcohols, 1608 cm^{-1} that corresponds with N-H bending vibration originated from amines group or amide I band, 2368 cm^{-1} arise from O-H stretching vibration of carboxylic acid, and 3390 cm^{-1} associated to N-H stretching vibration from amines group [3, 39]. The presence of these peaks gives information of the possibility that the carbonyl group formed amino acid residues, in which these residues act as a stabilizer that capped the AgNPs to prevent agglomeration [39, 40]. The absorption peak at 1384 cm^{-1} might be originated from C-N stretching vibration of the aromatic amide group [8, 39, 41], or N-H bond of primary and secondary amides [4], or C=C stretching vibration from amines [2] or N=O bending of nitro groups [5]. Absorption peaks between 779 to 619 cm^{-1} were assigned to O-H, where H-bonded with broad C-O of alcohols and phenols [4]. The results from FTIR spectra suggest that the synthesized AgNPs exhibited the presence of several functional groups that act as potential reducing agents, such as alkane groups, amine groups, and carboxylic acid.

3.3. XRD and EDS analysis of PCE and AgNPs

XRD characterization was conducted to investigate the crystalline properties of PCE and AgNPs that were synthesized from PCE. PCE shows an amorphous structure with a non-Bragg peak at the 2θ ranges of (15 – 25) °. The AgNPs synthesized from PCE in the different solvents show the emergence of crystalline peaks with different intensities and crystallinities (Figure 4).

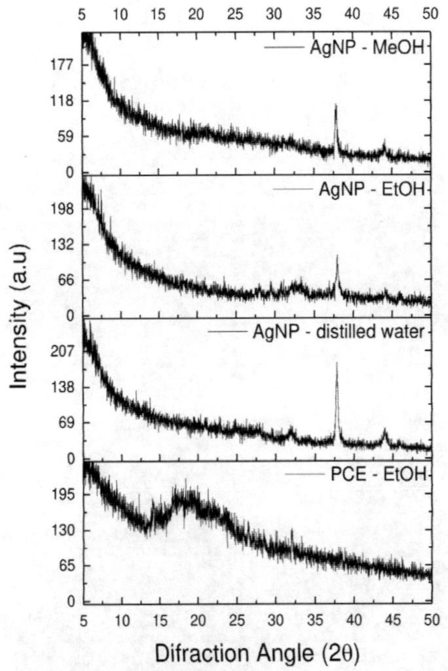

Figure 4. The XRD pattern of PCE in ethanol (E_0) and AgNP in distilled water (A_2), ethanol (E_2), and methanol (M_2) solution.

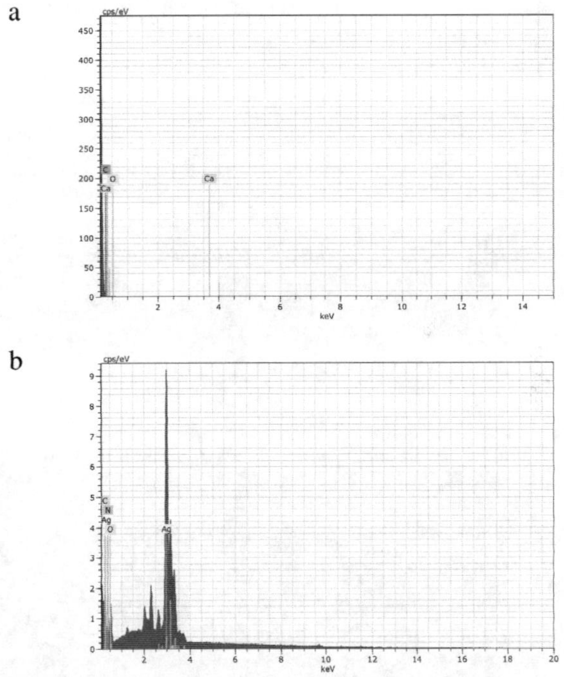

Figure 5. The EDS spectrum of PCE (a) and AgNPs in ethanol (b).

The crystallinity of the AgNPs was estimated by calculating the ratio of the integrated area of the crystallite peak with the integrated area of the entire XRD spectrum [42]. We found that the crystallinity was very low, specifically 3.0%, 1.1%, and 1.6% for AgNPs in distilled water (A_2), ethanol (E_2), and methanol (M_2) solution, respectively. The observed crystallite peak at 2θ position of 37.9° was originated from Ag (identified through powder diffraction file No 04-0783) with crystallite sizes of 31.42 nm, 50.27 nm, and 31.42 nm for A_2, E_2, and M_2, respectively, that was estimated using Scherer formula [43]. The Ag peak in E_2 was confirmed by the EDS result (Figure 5 and Table 1). It was detected as much as 17.49% of Ag atoms.

Table 1. Tabulated percentage of constituent atoms in PCE (E_0) and AgNPs (E_2) in ethanol.

Constituent atoms	Atomic Number	Percentage (%)	
		E_0	E_2
O	8	96.23	44.35
Ca	20	3.77	-
Ag	47	-	17.49
C	6	-	22.50
N	7	-	15.66

To corroborate the formation of nanoparticles, we provided the TEM results for the AgNPs solution where the silver nitrate dissolved in distilled water before being mixed with PCE extracted in ethanol (E_1), and AgNPs solution where both the silver nitrate and the PCE was dissolved in ethanol (E_2). The concentration of the silver nitrate in both samples is 0.1M. The results are shown in Figure 6 (a and b).

a.
b.

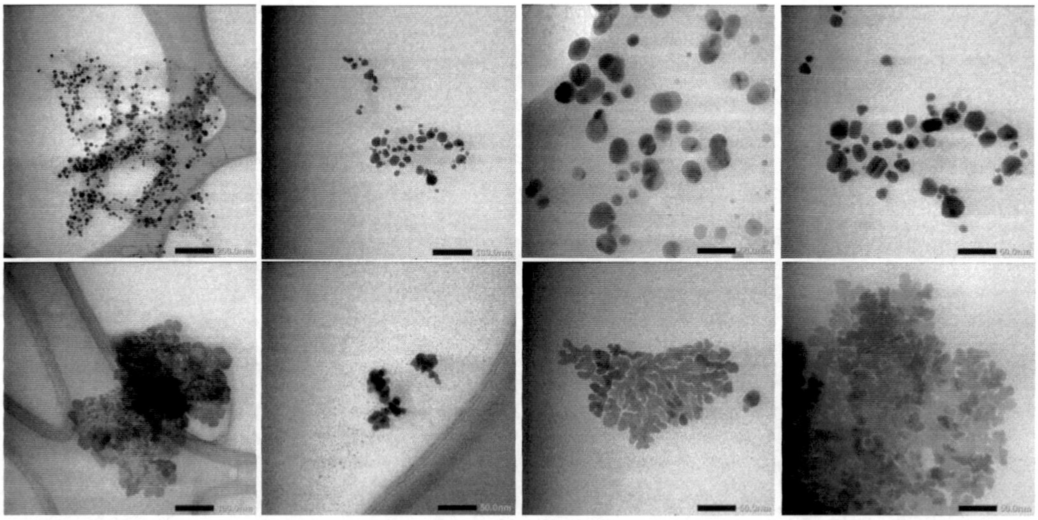

Figure 6. TEM results of AgNPs solution (a) where the silver nitrate dissolved in distilled water before mixed with PCE extracted in ethanol (E_1) and (b) AgNPs solution where both the silver nitrate and the PCE was dissolved in ethanol (E_2)

The AgNPs with sizes below 50 nm has been formed. There are differences in the shape and distribution of the AgNPs due to differences in the solvent that was used to dissolve silver nitrate ($AgNO_3$). The E_1 sample (Figure 6a) shows non-uniform spherical shapes with different sizes, whereas the E_2 sample (figure 6b) indicates an agglomeration, and the nanoparticles show irregular shapes.

3.4. LSV analysis of TiO_2 electrode immersed in PCE and AgNPs solution
LSV was affected by the reduction-oxidation (redox) reaction intensity going on in the working electrode. These reactions involve electron transfer and hence will result in voltage differences. The voltage difference converted to current. The LSV measurement results are shown in Figure 7 (a-d). The starting current of the TiO_2 electrode immersed in the AgNPs solution was always higher than the one that was immersed in PCE solution for all solvents, both in the dark and under light irradiation conditions (Figure 7a-7c). An anodic peak indicating reduction reaction [44] or reduction current (dark and light irradiated) was also observed. The reduction current takes place at a potential that was assigned as reduction potential (E_p). The reduction potential (E_p) provides information on the nature of the analyte involved [45]. The maximum reduction current or peak height (i_p) gives information on the analyte concentration, the number of electrons involved in the electrochemical process, and the possible presence of coupled chemical reactions [45]. The emergence of reduction peak indicates that there is a Faradaic current which is a sign of the occurrence of an electrochemical reaction. The reduction peak arises from the competition of the increasing rate of the electron transfer as the potential becomes more and more preferable for its occurrence with the increased thickness of the diffusion layer caused by the depletion of the depolarizer when the distance is more distant from the electrode surface [45].

The TiO_2 electrode immersed in A_0 and E_0 measured at light irradiated condition showed reduction current (i_p) peaked at the potential of -0.75 V and -0.77 V, respectively, while no reduction current for the TiO_2 electrode immersed in M_0 (Figure 7a-7d). The i_p for the TiO_2 electrode immersed in A_0 was 169.2% higher than the TiO_2 electrode immersed in E_0. The E_p significantly reduced, and its position shifted to a positive voltage (0.2 V) when the TiO_2 electrode was immersed in AgNP solution for All solvents (A_1, E_2, and M_2) in both dark and light irradiated conditions (Figure 7d). The i_p of TiO_2 electrode that was immersed in AgNPs, decreased as much as 72.5% and 46.1% from A_0 to A_1, and E_0 to E_2, in consecutive order. As for the TiO_2 electrode that was immersed in the M_2, the i_p is 0.08 mA/cm^2. The i_p in the light irradiation condition is smaller than the i_p measured in the dark.

The reduction potential indicates the tendency of a species to gain electrons and get reduced [46]. Thus, the shifted of reduction potential from -0.8 V in TiO_2 electrode that was immersed PCEs to +0.2 V in TiO_2 electrode that was immersed AgNPs shows that TiO_2 electrodes were immersed in AgNPs have more tendency to get reduced and also an indication of decreased in the resistance.

The emergence of a reduction peak was associated with the Faradaic process. This process related to an electrochemical process in the system that arise from (i) transfer of both the reactant from the solution bulk to the electrode surface and the product in the opposite direction, (ii) charge transfer reaction, (iii) possible structural rearrangement, and chemical reactions associated with charge transfer process [45]. We suspect that this electron transfer process might be due to the recombination process of electrons in the TiO_2 conduction band to the energy level of the electrolyte solution. In DSSC, the recombination process will lower the current density and hence the efficiency of the cell. Thus, in this case, the use of methanol as a solvent for purple cabbage extraction was recommended to minimize the recombination process (Figure 7c and 7d).

3.5. Optical properties characteristics of PCE and AgNPs
Figure 8 (a and b) shows the absorbance spectrum of PCE and AgNPs for all solvents. The extract shows absorption on the visible light range, peaked at (421, 480, 550, and 966) nm, (415 and 544) nm, and (409 and 544) for PCE in distilled water, ethanol, and methanol, respectively. The peak at (420 and 550) nm was a typical anthocyanin absorption peak [47]. The PCE extracted in distilled water shows more strong absorption compared to the PCE extracted in ethanol and methanol.

The peak around 544 nm to 550 nm originated from the aglycone group [48] in anthocyanin still existed in AgNPs that were synthesized using PCE that was extracted in ethanol (E_1) and methanol (M_1), while the silver nitrate was dissolved in distilled water (Figure 8b). This peak then disappeared in AgNP that was synthesized using PCE and silver nitrate that was both extracted and dissolved in distilled water (A_1), and AgNP that was synthesized using PCE, and silver nitrate that was dissolved in ethanol (E_2), and methanol (M_2), respectively (Figure 8). The peak at 973 still existed for all samples, but a new peak at (908 and 916) nm was observed in E_2 and M_2.

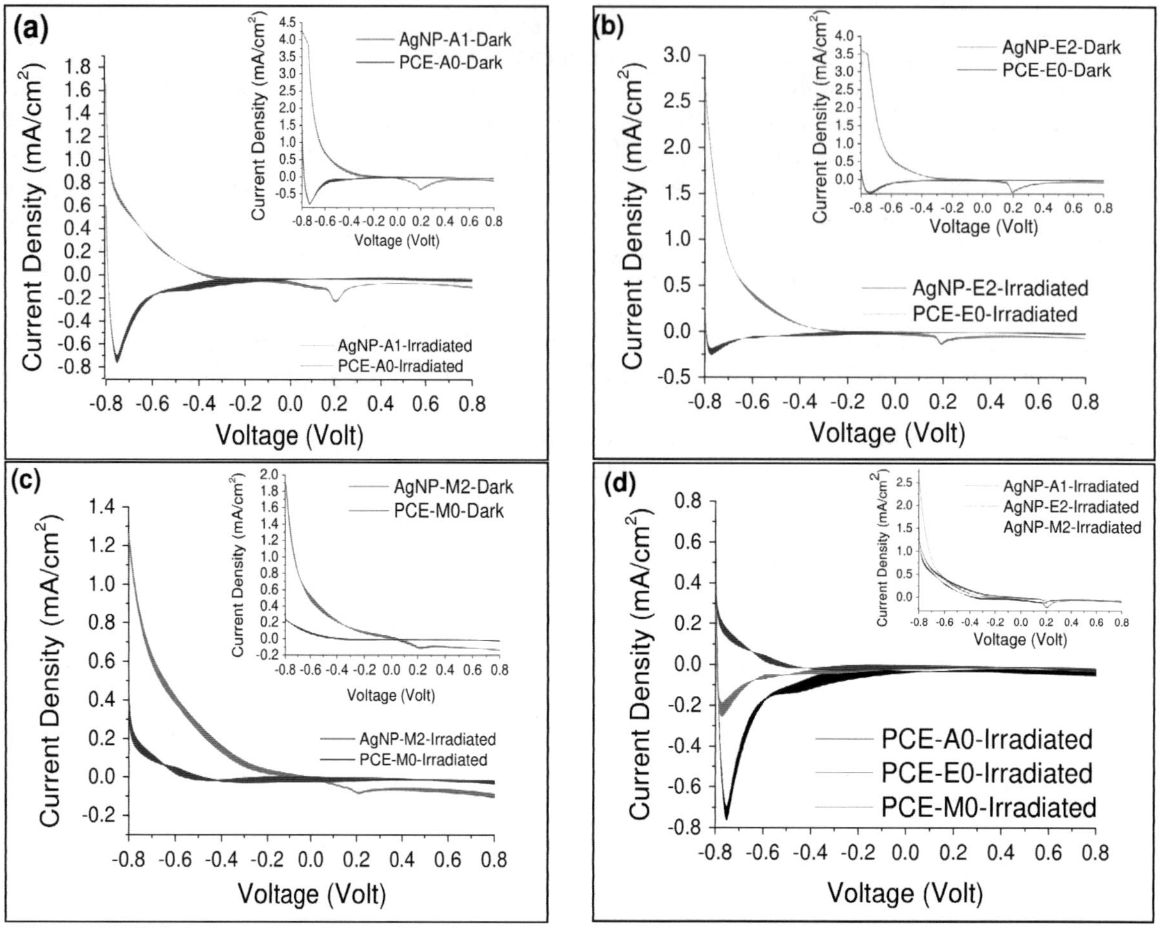

Figure 7. Voltammetry result of TiO_2 electrode immersed in PCE and AgNP synthesized in distilled water (a), ethanol (b), methanol (c) under light irradiation, and the comparison of PCE's and AgNP's voltammetry in all solvent (d). The inset in a, b, and c shows the voltammetry measured in dark conditions.

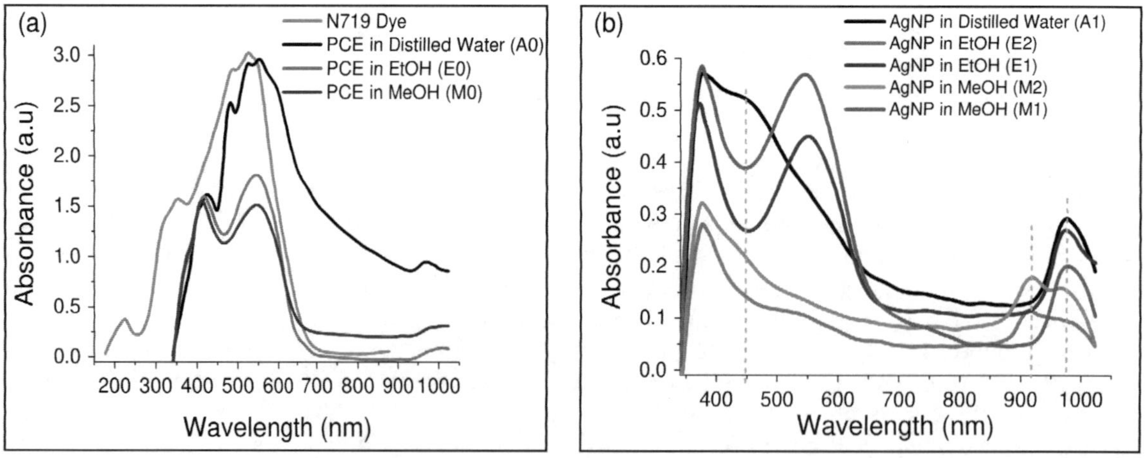

Figure 8. PCE's (a) and AgNPs' (b) absorbance spectrum in all solvents. The absorbance spectrum of N719 standard dye is shown as a comparison (a).

One can say that the utilization of different solvents for extracting purple cabbage and for dissolving silver nitrate is specious since no noticeable change in the absorbance spectrum (E_1 and M_1) in Figure 8b. The AgNPs absorbance peak (446 nm) was observed in A_1 (Figure 8b), however it was still overlapped with anthocyanin's absorption peak.

From the absorbance spectrum, we estimated the energy gap (E_g) value of the PCE and AgNPs by plotting photon energy ($h\nu$) versus absorbance coefficient squared (α^2). The highest photon energy in the plot is the energy [49]. The photon energy (E) and absorbance coefficient (α) was calculated using formula (1) and (2), respectively.

$$E = \frac{hc}{\lambda} \tag{1}$$

$$\alpha = \frac{A}{\lambda} \tag{2}$$

In which $h = 6.63 \times 10^{-34} Js$ is Planck's constant, $c = 3 \times 10^{8}\ m/s$ is the speed of light, λ is the wavelength, and A is the absorbance. The results are tabulated in Table 2. For all samples, the E_g were higher than the E_g for standard Ruthenium dye that was widely employed in the dye-sensitized solar cell specifically, 1.7 eV [50], 1.99 eV [51], and 2.29 eV [52] for N719 dye, even though theoretically, the reported E_g value of Ruthenium dye with different anchoring group can reach up to 2.65 eV [53].

Based on these results, we cannot conclude which one is better, smaller, or higher energy gap since we haven't utilized the PCE and AgNPs as the sensitizer in the DSSCs cell. Nevertheless, one requirement to be fulfilled is that no matter how big or small the E_g, the excited state level of the sensitizer or dye material has to be higher than the conduction band edge of the oxide semiconductor that was used as photo-anode (TiO_2 or ZnO) to ensure the efficient electron injection process [54]. Furthermore, the highest occupied molecular orbital (HOMO) of the dye materials should lie lower than the energy level of the redox electrolytes to maintain the dye regeneration process [54, 55].

Table 2. The energy gap value of PCE and AgNPs

Sample	Code	E_g (eV)
PCE	A_0	2.37
	E_0	3.00
	M_0	3.03
AgNPs ($AgNO_3$ dissolved in distilled water)	A_1	3.30
	E_1	3.36
	M_1	3.32
AgNPs ($AgNO_3$ dissolved in ethanol)	E_2	3.30
AgNPs ($AgNO_3$ dissolved in distilled methanol)	M_2	3.31

4. Conclusions

We have successfully extract crude anthocyanin from purple cabbage in three different solvents and the respected AgNPs synthesis processes. We identify that purple cabbage contains cyanidin with aglycone core. XRD diffractogram and TEM measurement confirm the existence of AgNPs however the crystallinity is very low, and we observe the AgNPs' absorbance peak at 446 nm in A_1. The PCE and

AgNPs contain carboxylic acid groups that originate from anthocyanin. This group ensures the attachment of both to the oxide semiconductor TiO_2. The dye attachment to the oxide semiconductor is a crucial factor to assure electron injection. The use of methanol as the solvent for purple cabbage extraction can eliminate the charge transfer processes due to the recombination process of electrons in the TiO_2 conduction band to the electrolyte.

Acknowledgments

This work is funded by DRPM DIKTI, through the Fundamental Research Scheme of 2021 with contract number of 112/E4.1/AK.04.PT/2021, and 18/UN29.20/PG/2021.

References

[1] Vert M, Doi Y, Hellwich KH, Hess M, Hodge P, Kubisa P, Rinaudo M and Schue F 2012 *Pure Appl. Chem.* **84(2)** 377-410

[2] Subramanian A and Sabesan M 2010 *Phcog Res.* **2(4)** 233-36

[3] Baskaran C and Ratha Bai C 2013 *Int. J. Drug Dev. & Res.* **5(1)** 1-10

[4] Koyyati R, Nagati V B, Nalvothula R, Merugu R, Kudle K R, Marx P and Padigya P R M 2014 *Int. J. Res. Pharm. Sci.* **5(1)** 32-9

[5] Paulkumar K, Gnanajobitha G, Vanaja M, Pavunraj M and Annadurai G 2017 *Adv. Nat. Sci.: nanosci. Nanotechnol.* **8** 035019

[6] Barros C H N, Cruz GCF, Mayrink W and Tasic Ljubica 2018 *Nanotechnol. Sci. Appl.* **11** 1-14

[7] Yaraki M T and Tan Y N 2020 *Chem. Asian. J* **15(20)** 3180-208

[8] Kurdekar A D, Chunduri L A A, Chelli S M, Haleyurgirisetty M K, Bulagonda E P, Zheng J, Hewlett I K and Kamisetti V 2017 *RSC Adv.* **7** 19863

[9] Eun-Ju K, Jae-Hwan K, Abdul-Majeed A and Yoon-Seok C 2011 *ACS Appl. Mater. Interfaces* **3** 1457-62

[10] Todescato F, Fortunati I, Minotto A, Signorini R, Jasieniak J J and Bozio R 2016 *Materials* **9(8)** 672

[11] Kyung W S, Jinhyun C, Aravindaraj G K, Sung L Y and Won K D 2016 *Sci. Rep.* **6** 26332

[12] Hong L V, Daeryeong B, Hun H J and Wonoh L 2018 *Compos Part B-Eng* **134** 141-50

[13] Khan I, Saeed K and Khan I 2019 *Arab. J. Chem* **12(7)** 908-31

[14] Jadoun S, Arif R, Jangid N K and Meena R K 2021 *Environ. Chem. Lett.* **19** 355–74

[15] Chokriwal A, Sharma MM and Singh A 2014 *Am. J. PharmTech. Res.* **4(6)**

[16] Vetchinkina E, Loshchinina E, Kupryashina M, Burov A and Nikitina V 2019 *Ind. Eng. Chem. Res.* **58(37)** 17207-218

[17] Ovais M, Khalil A T, Ayaz M, Ahmad I, Nethi SK and Mukherjee S 2018 *Int. J. Mol. Sci.* **19(12)** 4100

[18] Ovais M, Khalil A T and Islam N U 2018 *Appl Microbiol Biotechnol* **102** 6799-814

[19] Narayanan K B and Sakthivel N 2011 *Adv Colloid Interface Sci* **169** 59-79

[20] Narayanan K B and Sakthivel N 2010 *Adv. Colloid Interface Sci.* **156** 1-13

[21] Shankar S S, Rai A A and Sastry M 2004 *J Colloid Interface Sci.* **275** 496-502

[22] Manik U P, Nande A, Raut S and Dhoble S J 2020 *Result in Materials* **6** 100086

[23] Devanesan S and AlSalhi M S 2021 *Int J Nanomedicine* **16** 3343-56

[24] Amer M W and Awwad Akl M 2021 *Chemistry International* **7(1)** 1-8

[25] Yuan CG, Huo C, Gui B and Liu P 2017 *J Clust Sci* **28** 1319–33

[26] Rautela A, Rani J and Debnath (Das) M 2019 *J Anal Sci Technol* **10(5)**

Journal of Physics: Conference Series **2274** (2022) 012001 doi:10.1088/1742-6596/2274/1/012001

[27] Vankar P and Shukla D 2012 *Appl Nanosci* **2** 163-8

[28] Sarala E, Madhukara Naik M, Vinuth M, Rami Reddy Y V and Sujatha H R 2020 *J Mater Sci: Mater Electron* **31** 8589–96

[29] Demirbas A, Welt B A and Ocsoy I 2016 *Mat Lett* **179** 20-3

[30] Tamileswari R, Nisha M H and Jesurani Sr S 2015 *IJESRT* **4(7)**

[31] Chigurupati N, Saiki L, Gayser Jr C and Dash A K 2002 *Int J Pharm* **241** 293-99

[32] Gokilamani N, Muthukumarasamy N, Thambidurai M, Ranjita A and Velauthapillai D 2013 *J Sol-Gel Sci Technol* **66** 212-19

[33] Li Y, Ku S H, Chen S M, Ali M A and Alhemaid F M A 2013 *Int. J. Electrochem. Sci* **8** 1237-45

[34] Syafinar R, Gomesh N, Irwanto M, Fareq M and Irwan Y M 2015 *Energy Procedia* **79** 799-807

[35] Pratiwi D D, Nurosyid F, Supriyanto A and Suryana R 2017 *IOP. Conf. Ser.: Mater. Sci. Eng. (discontin)*

[36] Muchuweti M and Chikwambi Z 2008 *Am. J. Food Technol.* **3(1)**

[37] Favaro L I L, Balcao V M, Rocha L K H, Silva E C, Oliveira Jr J M, Vila M MDC and Tubino M 2018 *J. Braz. Chem. Soc* **29(10)** 2072-88

[38] Vasincu A, Paulsen B S, Diallo D, Vasincu I, Aprotosoaie A C, Bild V, Charalambous C, Constatinou A I, Miron A and Gavrilescu C M 2014 *Molecules* **19(11)** 19114-136

[39] Mathath R, Kadarkarai M, Chellasamy P, Jayapal S and Jiang-Shiou H 2013 *Parasitol Res* **112** 981-90

[40] Sathyavathi R, Balamurali K M, Venugopal R S, Saritha R and Narayana R D 2010 *Adv Sci Lett* **3** 1-6

[41] Mallikarjuna K, Narasimha G, Dillip G R, Praveen B, Shreedhar B, Lakshmi C S, Reddy B V S and Prasad Raju B D 2011 *Dig J Nanomat Biostruct* **6(1)** 181-86

[42] Jaworska M, Sakurai K, Gaudon P and Guibal E 2003 *Polym. Int.* **52** 198-205

[43] Arsyad Waode S, Januari A, Jumalia, Ilmawati W S, Usman I and Hidayat R *Journal of Non-linear Optical Physics & Materials* **27(2)** 1850019

[44] Thomas M and Rajiv S 2019 *New J. Chem* **43** 4444-54

[45] Bontempelli G, Dossi N and Toniolo R 2016 Voltammetry Linear Sweep and Cyclic in *Reference Module in Chemistry, Molecular Sciences and Chemical Engineering* (Italy, Elesevier) p 1-10

[46] Bhatt V 2016 Thermodynamics and kinetics of complex formation in *Essenstials of coordination chemistry* (Academic Press) p 111-37

[47] Jacquemoud S and Ustin S 2019 Spectroscopy of leaf molecules in *Leaf Optical Properties* (Cambridge University Press) p 48-73

[48] Lopes T J, Yaginuma S R, Quadri M G N and Quadri M B 2011 *Braz. Arch. Biol. Technol* **54(6)** 1349-56

[49] Oviri O K and Ekpunobi A J 2013 *Res. J. Recent Sci* **2(1)** 25-31

[50] Jaeun A, Ki Chang L, Dajung K, Chong Chan L, Sunae L, Dae Won C, Sukhun K and Chan I, 2013 *Mol. Cryst. Liq. Cryst.* **581(1)** 45-51

[51] Ashraf S, Akhtar J, Siddiqi H M and El-Shafei A 2017 *New J. Chem* **41** 6272-77

[52] Abboto A, Sauvage F, Barolo C, De Angelis F, Fantacci S, Graetzel M, Manfredi N, Marinzi C and Nazeeruddin MK 2011 *Dalton Trans* **40** 234-42

[53] Chitumalla R K and Jang J 2018 *Sol. Energy* **159** 283-90

[54] Carella A, Borbone F and Centore R 2018 *Front. Chem* **6** 481

[55] Sharma K, Sharma V and Sharma S S 2018 *Nanoscale Res. Lett.* **13(381)**

[56] Mahitha B, Deva Prasad R B, Dillip G R, Madhukar R C, Mallikarjuna K, Manoj I, Priyanka S, Jayantha R K and John S K 2011 *Dig J Nanomat Biostruct* **6(1)** 135-42

Journal of Physics: Conference Series **2274** (2022) 012002 doi:10.1088/1742-6596/2274/1/012002

The effect of addition of WS_2 nanosheets on the performance of ZnO nanorods-based photodetector

Aufa Salsabilla[1] , V. Fauzia[1*], and C. Imawan[1]

[1] Department of Physics, FMIPA, Universitas Indonesia, Depok, Indonesia.

[*]Corresponding author: vivi@sci.ui.ac.id

Abstract. Fabrication of the heterostructure of ZnO/transition metal dichalcogenides (TMD) has attracted much attention to improve the performance of ZnO-based photodetectors. In this work, we report the effect of the addition of WS_2 nanosheets on the performance of ZnO nanorods based photodetector*s.* ZnO nanorods were prepared using the hydrothermal method, and the few layers of WS_2 nanosheets were prepared by liquid-phase exfoliation. Exfoliated WS_2 solution was deposited on the surface of ZnO nanorods by spin-coating method, and the photodetector performance was calculated via current-voltage measurement in the dark and under laser light sources with a wavelength of UV 365, cyan 505, and red 625 nm. The results show that the dark current significantly decreased after the addition of WS_2 on the ZnO nanorods surface. The UV detector performances sensitivity is increased from 129% to 334%, responsivity is increased from 0.93 A/W to 1.11 A/W, and the detectivity is increased from 2.5×10^9 to 4.4×10^9 Jones. However, both samples cannot be used as visible light detectors because of the low photogenerated current.

1. Introduction

The photodetector is an essential part of light-sensing technology that can convert optical energy to electrical energy [1]. The photodetector is usually used in a UV radiation monitor, missile detector, fire alarm and other optical communication devices [1–3]. The primary material of the photodetector is a semiconductor that can absorb light to excite electrons that produce an electric current. [4,5]. Photo-sensitive materials prepared in nanostructures show higher performances compared with when it prepared in bulk structures [6][2]. ZnO is one of the semiconducting materials widely used as UV-sensitive materials. ZnO has a wide bandgap (3.37 eV) in hexagonal wurtzite structure and has excellent thermal and mechanical stability [7][8][9]. However, ZnO still has several limitations, such as a high recombination rate and a high dark current; hence the photodetector performance is not optimal. To improve the performances of the photodetector, we deposited semiconducting transition metal dichalcogenides (TMD) tungsten disulfides WS_2. WS_2 is a two-dimensional material which layers are weakly bounded by the Van der Waals bond [10]. It has a low bandgap that can change from indirect bandgap (1.3 eV) to direct and higher bandgap (~2 eV) when the thickness approaches monolayer [11]. Only limited studies of fabrication of UV photodetector based on ZnO/WS_2 heterostructure have been published. This work shows that by the addition of WS_2 nanosheets on the surface of ZnO nanorods, the performances of UV photodetector could be increased.

2. Material and Method

2.1. Materials
Zinc acetate dehydrates 0.2M ($C_4H_6O_4Zn$ $2H_2O$), *zinc nitrate tetrahydrate* 0.05M (N_2O_6Zn $4H_2O$), *hexamethylenetetramine* 0.05M ($C_6H_{12}N_4$) and *tungsten disulfide* powder (SKU 243639, 2 μm, 99%) were purchased from Merck.

2.2. Synthesis of ZnO Nanorods
Zinc acetate dehydrates of 0.2M was dissolved in 20 mL deionized water as a ZnO seed solution. The ZnO seed solution was sprayed on the interdigitated ITO (Osilla S161) using an ultrasonic nebulizer for 15 min heated on the hotplate at 450°C. Then, the ZnO seeding process was continued by heating at 450°C for 1 h. After that, the ZnO thin film coated substrates were soaked in 20 ml mixture of 0.05M *zinc nitrate tetrahydrate* and 0.05M *hexamethylenetetramine* and heated for 6 h at 95°C inside an oven [1].

2.3. Preparation of WS_2 Nanosheets
WS_2 nanosheets were prepared by liquid-phase exfoliation method where 3 mg/mL WS_2 bulk powder was dispersed in a mixture of 20% *isopropanol* and 80% deionized water. Then, the solution was sonicated at 270 W for 8 h with 6s ON and 4s OFF modes. After that, the solution was centrifugated twice with 14100 RCF for 3 min and 9 min to separate the exfoliated WS_2 nanosheets from the bulk one [12]. 100 μL exfoliated WS_2 nanosheets were then deposited on the surface of ZnO nanorods using the spin-coating method with 500 rpm for 10 s and 700 rpm for 15 s. After that, the sample was annealed for 45 min on the hotplate at a temperature of 60°C.

2.4. Characterization
The morphological, structural and optical properties of ZnO/WS_2 were observed at room temperature by field emission electron microscope (FESEM; Zeiss Supra 55 V P), transmission electron microscope (TEM; FEI 200 kV Tecnai G20 S Twin), WITec Raman microscope with a 532 nm laser source and Thermo Fisher Scientific GENESYS 10S. The current-voltage (I-V) measurements were recorded by Keithley 2400 electrometer in the dark and under laser light sources (Thorlabs) with a wavelength of 365, 505 and 625 nm with an optical power of 3 mW, as shown in Fig. 1.

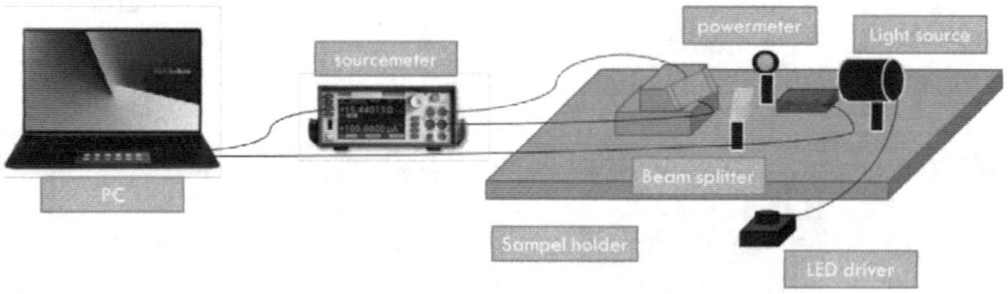

FIGURE 1. Photodetector characterization Set-Up

3. Results and Discussions
The surface morphology of the heterostructures ZnO/WS_2 can be seen from the FESEM image, as shown in Figure 2(a). ZnO nanorods grew perpendicular to the interdigitated ITO substrate with a diameter from 76 nm to 206 nm. WS_2 nanosheets were generally partially covered the surface of ZnO nanorods as indicated by a rougher surface of ZnO. The TEM image in Figure 2(b) shows the number layer of WS_2 with an interlayer distance of 0.62 nm.

The Raman spectrum in Figure. 3(a) show 2 peaks at 324 cm^{-1} and 421 cm^{-1}, indicating E^1_{2g} and A_{1g} peaks of WS_2, respectively. The distance between the two peaks is 67 cm^{-1}. This shows that a few layers of WS_2 have been formed, which are consistent with the TEM observation in Figure 2(b). [5,13].

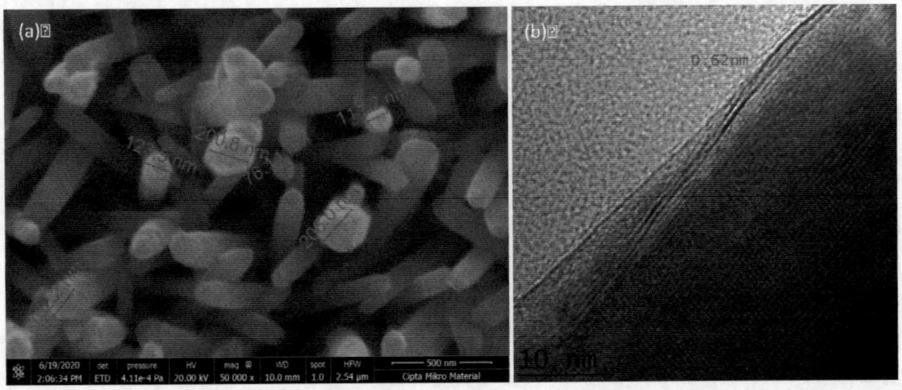

FIGURE 2. (a) FESEM and (b) TEM images of ZnO/WS_2 heterostructure

We further measured the absorption spectra of WS_2 nanosheets solution, ZnO nanorods, and ZnO/WS_2 heterostructure using a UV-Vis spectrometer. The absorbance spectrum of WS_2 nanosheets solution in Figure 3(b) shows there are 3 peaks at 415 nm, 517 nm, and 623 nm and then used to calculate the energy band gap (2.48 eV) as shown in the Tauc plot in Figure 3(c). Figure 3(d) shows the absorbance spectrum of pristine ZnO and ZnO/WS_2 heterostructure. There are no absorption spectrum changes after the addition of WS_2 nanosheets on the ZnO nanorods surface. This feature may be due to the small amount of deposited WS_2; hence the energy band gap of ZnO nanorods and ZnO/WS_2 heterostructure are similar to 3.19 eV as shown in the Tauc plot in Figure 3(e) and 3(f).

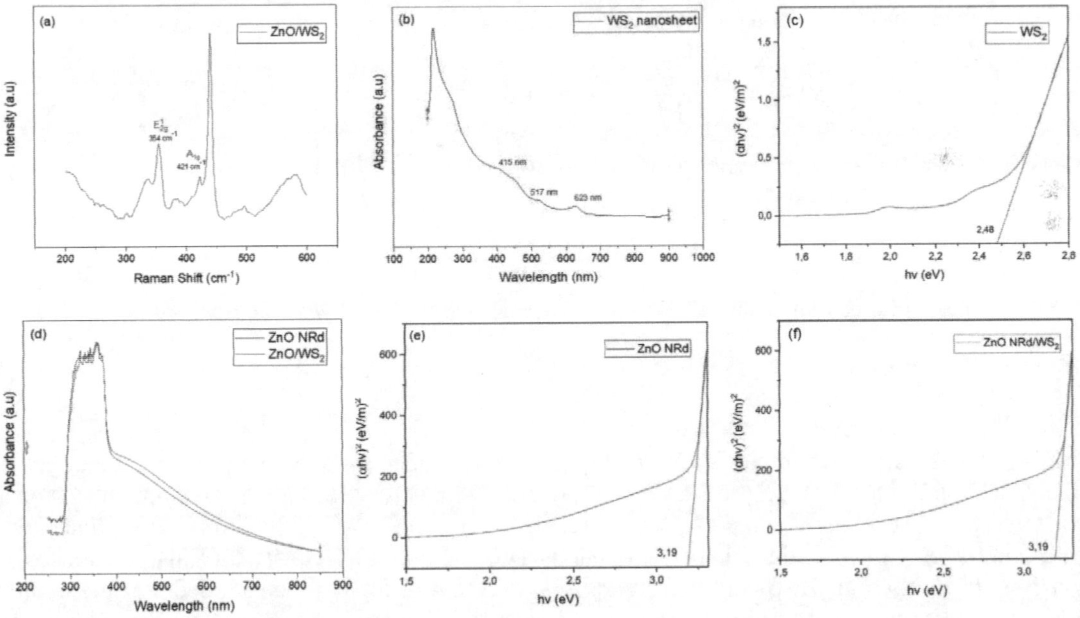

FIGURE 3. (a) Raman spectra of ZnO/WS_2, (b) the UV-Vis absorbance spectrum and (c) Tauc plot of WS_2 nanosheets solution,(d) the UV-Vis absorbance spectrum of ZnO/WS_2 and Tauc plots of (e) ZnO nanorods and (f) ZnO/WS_2

Optoelectrical properties of the photodetector were characterized through I-V measurement in the dark and under laser illumination, as shown in Figure 4. Figure 4(a-c) shows the dark current and photocurrent of ZnO/WS_2 are decreased compared to the pristine ZnO under a bias voltage of more than 3 V. Dark current can arise from ZnO crystal defects which produce free electrons that can move in dark

Journal of Physics: Conference Series **2274** (2022) 012002 doi:10.1088/1742-6596/2274/1/012002

conditions while the photocurrent comes from the photogenerated electrons and holes. Figure 4 also shows that both the pristine ZnO and ZnO/WS_2 heterostructure produce higher photocurrent under UV light (365 nm) compared to those under cyan (505 nm) and red (625) light sources. These results reveal that ZnO works better for UV-light detectors than cyan and red-light detector due to ZnO nanorods having a wide bandgap (3.19 eV), as presented in Figure 3(f). Decreasing dark current and photocurrent may be due to the surface passivation effect of ZnO by WS_2 nanosheets. The active sites of the WS_2 nanosheets on the ZnO surface may be able to trap these electrons, thereby reducing the current, especially when high voltage is applied above 3 V [14].

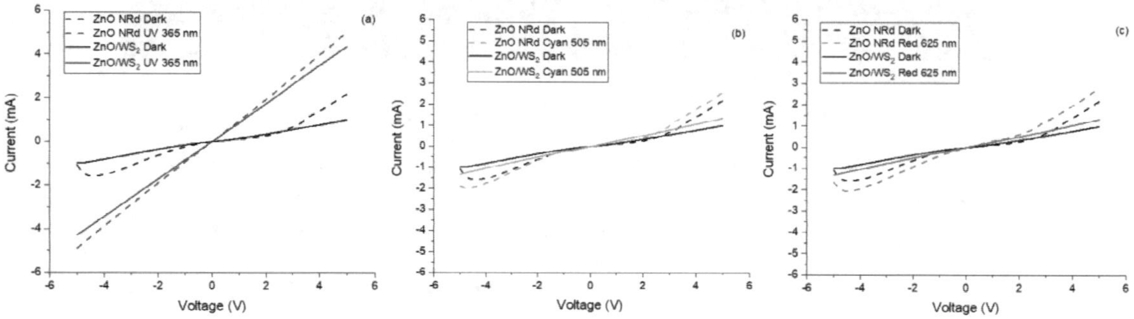

FIGURE 4. I-V curves of ZnO and ZnO/WS_2 heterostructures in the dark and under (b) UV light 365 nm; (c) Cyan light 505 nm; (d) Red light 625 nm.

Table. 1 shows the comparison of ZnO and ZnO/WS_2 photodetector performance consisting of sensitivity, responsivity, and detectivity measured at 5V applied bias voltage. Sensitivity is the ratio between photocurrent (I_{light}) and dark current(I_{dark}).

$$(S\,\%) = \frac{I\,light - I\,dark}{I\,dark}\ x\ 100\% \tag{1}$$

Responsivity is a ratio between the net photocurrent to the power of light.

$$R_\lambda = \frac{I_{light} - I_{dark}}{P} \quad (A/W) \tag{2}$$

Detectivity is a parameter to measure the ability of the device to detect weak optical signals follows.

$$D^* = \frac{R_\lambda\, A^{1/2}}{\sqrt{2}e\sqrt{I_{dark}}} \quad (Jones) \tag{3}$$

Table 1 shows that both samples show the highest performance as UV detectors, and it can be said that both samples could not be used as cyan and red light detectors due to the poor performance parameter values. At a bias voltage of 5V, the sensitivity of ZnO/WS_2 (334%) is higher than that of pristine ZnO (129%). The addition of WS_2 nanosheets on the surface of ZnO also increases the responsivity from 0.938 to 1.116 and detectivity from 2.5 to 4.4. It can be concluded that the addition of WS_2 nanosheets is useful to improve the ZnO-based UV detector by suppressing the dark current significantly.

TABLE 1. Photodetector Performances

Lights		Sensitivity (%)	Responsivity (A/W)	Detectivity (10^9 Jones)
UV	ZnO	129	0.938	2.5
	ZnO/WS_2	334	1.116	4.4
Cyan	ZnO	16	0.117	0.3
	ZnO/WS_2	33	0.111	0.4
Red	ZnO	21	0.199	5.3
	ZnO/WS2	33	0.109	0.4

4. Conclusions

In this work, we have studied the effect of the addition of WS_2 *nanosheets* on the performance of photodetectors based on ZnO *nanorods*. The result shows that the WS_2 *nanosheets* deposited on ZnO *nanorods* using the spin-coating method is effectively able to enhance the performance of the UV detector where the sensitivity increased from 129% to 334%, responsivity increased from 0.93 A/W to 1.11 A/W, and the detectivity increased from 2.5 x 10^9 Jones to 4.4 x 10^9 Jones. The WS_2 addition can suppress the dark current of ZnO, where the WS_2 layer plays a role as an electron trap center. However, ZnO or ZnO/WS_2 based photodetector cannot be used as cyan and red photodetector because of the low photogenerated current.

Acknowledgments

This research was supported by the Department of Physics, Faculty of Mathematics and Sciences, Universitas Indonesia.

References

[1] Nguyen T T, Patel M, Ban D K and Kim J 2018 Vertically trigonal WS 2 layer embedded heterostructure for enhanced ultraviolet–visible photodetector *J. Alloys Compd.* **768** 143–9

[2] Tsai S-H, Shen Y-C, Huang C-Y and Horng R-H 2019 Deep-ultraviolet Schottky photodetectors with high deep-ultraviolet/visible rejection based on a ZnGa2O4 thin film *Appl. Surf. Sci.* **496** 143670

[3] Xie Y, Wei L, Wei G, Li Q, Wang D, Chen Y, Yan S, Liu G, Mei L and Jiao J 2013 A self-powered UV photodetector based on TiO2 nanorod arrays *Nanoscale Res. Lett.* **8** 1–6

[4] Zheng M, Gui P, Wang X, Zhang G, Wan J and Zhang H 2019 Applied Surface Science ZnO ultraviolet photodetectors with an extremely high detectivity and short response time *Appl. Surf. Sci.* **481** 437–42

[5] Perea-Lõpez N, Elías A L, Berkdemir A, Castro-Beltran A, Gutiérrez H R, Feng S, Lv R, Hayashi T, Lõpez-Urías F, Ghosh S, Muchharla B, Talapatra S, Terrones H and Terrones M 2013 Photosensor device based on few-layered WS2 films *Adv. Funct. Mater.* **23** 5511–7

[6] Zhang Y, Ram M K, Stefanakos E K and Goswami D Y 2012 Synthesis, characterization, and applications of ZnO nanowires *J. Nanomater.* **2012**

[7] Shewale P S, Lee S H and Yu Y S 2018 UV sensitive pulsed laser deposited ZnO thin films: Influence of growth temperature *J. Alloys Compd.* **744** 849–58

[8] Özgür Ü, Alivov Y I, Liu C, Teke A, Reshchikov M A, Doğan S, Avrutin V, Cho S J and Morkọ H 2005 A comprehensive review of ZnO materials and devices *J. Appl. Phys.* **98** 1–103

[9] Jin S E and Jin H E 2019 Synthesis, characterization, and three-dimensional structure generation of zinc oxide-based nanomedicine for biomedical applications *Pharmaceutics* **11**

[10] Han J J, Zhang Q Y, Huang M Y, Chen Y, Yan X and Lang W Z 2020 Two-dimensional WS2 membranes constructed on different substrates for efficient dye desalination *Desalination* **480**

[11] Zhao H, Wu H, Wu J, Li J, Wang Y, Zhang Y and Liu H 2019 Preparation of MoS2/WS2

nanosheets by liquid phase exfoliation with assistance of epigallocatechin gallate and study as an additive for high-performance lithium-sulfur batteries *J. Colloid Interface Sci.* **552** 554–62

[12] Dwiputra M A, Fadhila F, Imawan C and Fauzia V 2020 The enhanced performance of capacitive-type humidity sensors based on ZnO nanorods/WS2 nanosheets heterostructure *Sensors Actuators, B Chem.* **310** 127810

[13] Butanovs E, Vlassov S, Kuzmin A, Piskunov S, Butikova J and Polyakov B 2018 Fast-Response Single-Nanowire Photodetector Based on ZnO/WS2 Core/Shell Heterostructures *ACS Appl. Mater. Interfaces*

[14] Lan C, Li C, Wang S, Yin Y, Guo H, Liu N and Liu Y 2016 ZnO-WS2 heterostructures for enhanced ultra-violet photodetectors *RSC Adv.* **6** 67520–4

Growth of silver nanoparticles on unpolished silicon substrate by used of microwave-assisted technique to enhance the Raman spectrum of glucose

Isnaeni*, Nurfina Yudasari, and Affi Nur Hidayah

Research Center for Physics, National Research and Innovation Agency, Indonesia, Building 440-442, Puspiptek, South Tangerang, Banten, Indonesia 15314

*Corresponding author: isnaeni@brin.go.id

Abstract. Silver nanoparticle is useful metallic nanoparticle due to plasmonic property. One of the most important application of silver nanoparticle is enhancing Raman spectrum using so-called surface-enhanced Raman spectroscopy (SERS). Mostly, researchers synthesize colloidal silver nanoparticles and deposite on substrate to form SERS substrates. In this work, we simplified the synthesis of SERS substrate by use of microwave. We have successfully grown silver nanoparticles on unpolished silicon substrate using single step microwave-assisted technique. The unpolished silicon substrate has micro-size chambers that are useful for Raman measurement. The silver nanoparticles on silicon substrate were approximately 88 nm in diameter. Our experiment using corn syrup revealed that our SERS substrate enhance Raman peaks spectra of glucose and galactose by more than 16 times.

1. Introduction

Silver nanoparticle is one type of the most promising nanoparticles. It is not only due to nano-scale size of the particle that delivers new physical and chemical properties, but also the simplicity in the synthesis process of the colloid [1]. Silver nanoparticles have different property in comparison to that of silver bulk that is called localized surface plasmon resonance (LSPR) [1,2]. Plasmon in metallic structures is a collective oscillations of their free electrons in response to incident waves [3,4]. This plasmonic property can enhance surrounding photons or electromagnetic fields [5]. One of the promising application of silver nanoparticles is surface-enhanced Raman spectroscopy (SERS) [6].

Enhancement of Raman signal due to silver nanoparticles is strongly related to plasmonic property of silver nanoparticles [2]. Plasmon generates high density electromagnetic field surrounding silver nanoparticles. This field may contribute to enhance the Raman signal via energy transfer between electromagnetic field and vibrational molecules [3]. The electromagnetic field is found to increase in between two silver nanoparticles that creates so-called hot spot area. This hot spot has enormous field enhancement that can enhance Raman signal for tens to hundreds times [7]. In the practical application, silver nanoparticles are coated on the substrate, which is known as SERS substrate. Mostly, researchers start to prepare good colloidal silver nanoparticles and then to deposite them on a certain substrate using spin coating technique [8], Langmuir Blodgett technique [9], drop dry technique [2], block copolymer process [10] and printing technique [11] which are known quite complicated. Furthermore, the silver nanoparticles are sometimes not easily to deposit on the substrate, thus the silver nanoparticles disappear when the substrate is immersed in liquid sample or analyte. Therefore, we need a simple and fast technique to make deposition of silver nanoparticles on substrate with good strength of deposition.

The purpose of this work is to fabricate silver nanoparticles on a substrate using one step microwave-assisted technique for SERS application. We limited our work on unpolished silicon substrate and corn syrup contained glucose as analyte, in order to enhance the Raman spectroscopy investigation.

2. Experimental Methods

Synthesis of silver nanoparticles was done by prepared silver nitrate, ethyl glycol (Merck) and polyvinylpyrrolidone (VWR Life Science). Single side polished silicon wafer was cut to small size approximately 1 cm × 1 cm. The silicon substrate was cleaned using ethanol and distilled water. Silver nitrate (100 mg) and polyvinylpyrrolidone (160 mg) were dissolved in ethyl glycol (100 ml) in a glass beaker and were stirred using magnetic stirrer at 500 rpm for 30 minutes. For microwave process, 2 ml of mixed solution was placed in a small 5 mL glass beaker. The silicon substrate was inserted and immersed in the beaker. The beaker was placed in the domestic microwave and heated for 45 seconds at power 800 W as shown in Figure 1. The solution and substrate was then naturally cooled down. The substrate contained silver nanoparticles was then taken out from the beaker and washed using distilled water for several times. The remaining silver nanoparticles solution in the small beaker was then separated for UV-Vis spectrometry measurement.

Figure 1. One-step process for growing silver nanoparticles on silicon substrate.

Unpolished silicon substrate with and without silver nanoparticles was then observed using 3D-digital microscope Keyence VHX-700. In order to observe the size of silver nanoparticles, we used field-emission scanning electron microscope (FE-SEM) JEOL JIB-4610F. For Raman spectroscopy characterization, we placed a small drop of corn syrup on the unpolished silicon substrate and unpolished silicon substrate with silver nanoparticles. The Raman spectra were obtained using Horiba Portable Raman with a 10× objective lens, iHR-320 spectrometer and excitation laser at 532 nm wavelength. We also utilized homemade UV-Vis spectrometer for measuring absorbance using MAYA Pro2000 spectrometer.

3. Results and Discussions

In our experiment, the growth of silver nanoparticles on silicon substrate were successfully conducted. Silver nanoparticles were deposited on polished or unpolished side of the silicon substrate. However, we only used unpolished side of substrate for this work. As the result, we observed grey colour appearance of unpolished silicon substrate without silver nanoparticles (inset of Figure 2). On the other hand, we observed green-yellowish colour appearance of unpolished silicon substrate with silver nanoparticles (inset of Figure 3). The colour change indicates that silver nanoparticles are formed a coating layer on the silicon substrate. Under 3D-microscope, we observed dark-blue colour from unpolished silicon substrate (Figure 2). On the other hand, we noticed dark-purple colour from surface of unpolished silicon substrate contained silver nanoparticles (Figure 3). The different colour of silver nanoparticles on unpolished silicon substrate as observed by naked eyes and 3D-microscope was probably due to different angle of incident light and observation. Nevertheless, the colour change after silver nanoparticle deposition suggested that thin layer of silver nanoparticles has been grown and deposited on unpolished sides of silicon substrate. We also noticed that the deposited silver nanoparticles were firmly attached to the surface of silicon substrate, eventhough we had done several time of washing of the sample using distilled water but the silver nanoparticles still remained on the surface. This indicated that we had synthesized strong layer of SERS substrate.

Figure 2. Unpolished silicon substrate without silver nanoparticles.

Figure 3. Unpolished silicon substrate with silver nanoparticles.

The unpolished side of silicon substrate gives an advantage for micro Raman SERS. As shown in Figure 4, the unpolished side of silicon substrate had random voids or chambers. The shape of the chamber was random rectangular. This micron size chamber had dimension approximately 25 µm × 25 µm, with the depth approximately 3.5 µm as shown in the inset of Figure 4. Number of micro-size chambers on the surface were enormous all over the surface. These micron-size chambers are useful for Raman measurement, especially liquid samples or analyte. The chambers are able to maintain small amount of the liquid sample inside the chambers during Raman measurement. This is done to make sure that analyte is in good contact with silver nanoparticles to get Raman enhancement. On the other hand, flat surface of SERS substrate sometimes could not maintain sample in place, since liquid sample tends to spread out. Therefore, utilizing such micro-size chambers are beneficial for further Raman measurement.

Figure 4. Micro-size chamber on unpolished silicon substrate

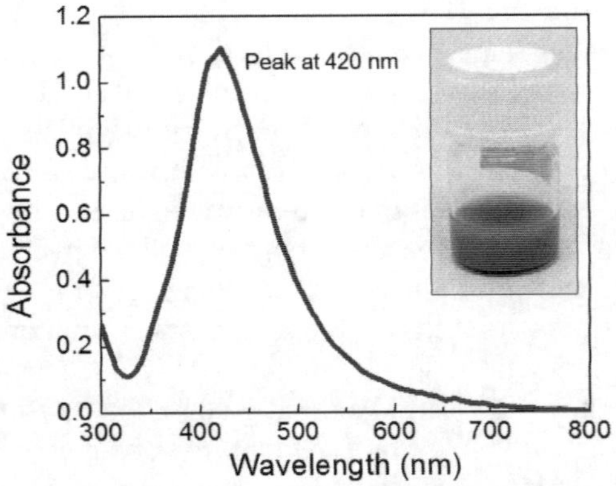

Figure 5. Absorbance of leftover colloidal silver nanoparticles.

Our synthesized silver nanoparticles in solution as shown inset Figure 5 was characterized by UV-Vis spectrometer. The liquid of silver nanoparticle had dark brown colour. This indicates that the size of silver nanoparticles are around 50 nm until 200 nm. From absorbance spectra we also noticed that the absorbance peak wavelength, which indicates plasmonic peak, was at 420 nm wavelength. In usual, the small size of silver nanoparticles (below 10 nm in diameter) show the plasmonic peak at 400 nm of wavelength [1,12]. In our case, the silver nanoparticles are clearly visible using 3D-microscope image as shown in Figure 3 and 4. We also used FE-SEM in order to confirm the size of the silver nanoparticles on the unpolished silicon substrate. From FE-SEM image in Figure 6, it is clearly seen that silver nanoparticles grow on the surface and their size are quite random. Histogram in Figure 7 shows that silver nanoparticles has a distribution of size in the range of 30 nm until 160 nm. Using Gaussian fitting we can conclude that average size of silver nanoparticles is approximately 88 nm. Thus, our FE-SEM image result is agree with absorbance spectrum and colour appearance of silver nanoparticles. In addition, the inset of Figure 6 showed the micro-size chambers on the surface of the unpolished silicon substrate.

Figure 6. FE-SEM image of silver nanoparticles on the unpolished silicon substrate.

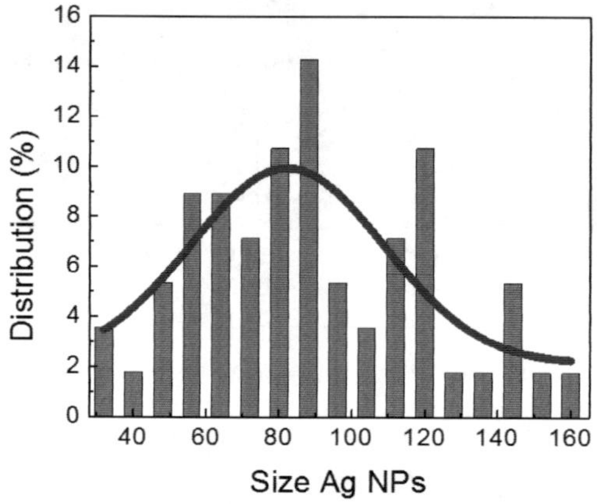

Figure 7. Size distribution of silver nanoparticles obtained from FE-SEM.

Raman measurement was done using corn syrup as analyte and it placed on the unpolished silicon substrate with or without silver nanoparticles. The same measurement parameters were used for both samples to ensure valid comparison, such as acquisition time, magnification, object lens in the microscope, excitation laser power and wavelength. The amount of corn syrup dropped on each substrate was approximately 2 μL. Figure 8 shows the Raman spectra of corn syrup on unpolished silicon substrate with and without silver nanoparticles. It is clearly seen that without silver nanoparticles, the Raman spectrum are weak (the blue line in Figure 8). Some peaks are overlapped so that broad peaks occur. On the other hand, we also noticed that Raman spectrum of corn syrup with silver nanoparticles on the unpolished silicon substrate is enhanced significantly. The Raman intensity increase and it become sharp and clear. The Raman enhancement of glucose signal originates from the interaction between glucose and silver nanoparticles. In our case, there was no special treatment of glucose prior deposition, thus the glucose molecules that naturally attached to the silver nanoparticles received plasmon effect from silver nanoparticles. The plasmonic effect of silver nanoparticles gave an additional density of states that matches with Raman vibrational energy of glucose.

We noticed that the Raman spectrum consists of many peaks which are correspond to vibrational frequency of molecules in the corn syrup. The dominant molecule in corn syrup is glucose and other is galactose. We had identified glucose and galactose peaks as shown in Figure 8 using symbol G and T, respectively. We observed that most of glucose peaks are observed in Raman spectrum of corn syrup without silver nanoparticles. However, most of galactose molecule peaks are not visible on silicon substrate without silver nanoparticles. These Raman peaks strongly occur on the unpolished silicon substrate with silver nanoparticles.

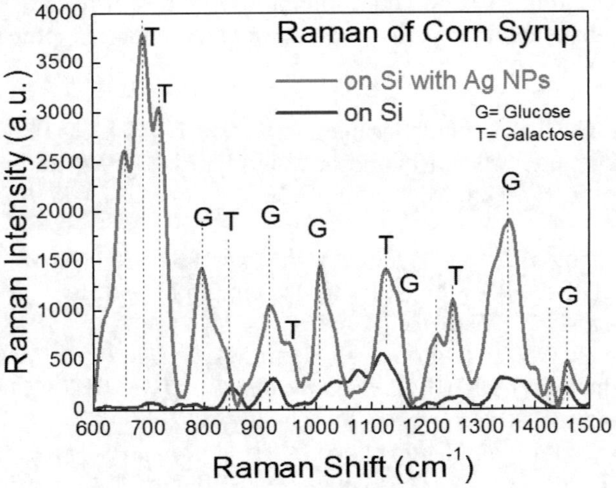

Figure 8. Raman spectra of corn syrup on the unpolished silicon substrate with and without silver nanoparticles.

In Table 1 and Table 2, we calculated Raman enhancement factor for glucose and galactose from Figure 8. We identified six Raman peaks of glucose and we found that by addition of silver nanoparticles on silicon substrate, the Raman intensity increases by 2.3 times to 68 times, with average enhancement was 16 times. In addition, we also identified galactose peaks with average enhancement 26 times by addition of silver nanoparticles. We notice that enhancement factor for each Raman peaks are not the same because some energies of vibrational molecules resonate with plasmonic energy of silver nanoparticles, even some other vibrational energies are not resonance with plasmonic energy. Therefore, the enhancement factor for each peak is not the same. We also notice that enhancement factor is high at low Raman shift, which is close to excitation laser for Raman and plasmonic energy of silver nanoparticles.

Table 1. Raman peaks of glucose.

Raman Shift (cm^{-1})	Enhancement	Molecule
797.7	68.2	Glucose
953.7	13.9	Glucose
1037.7	2.5	Glucose
1127.9	2.6	Glucose
1353.1	6.3	Glucose
1460.3	2.3	Glucose

Table 2. Raman peaks of galactose.

Raman Shift (cm^{-1})	Enhancement	Molecule
691.2	69.3	Galactose
720.5	64.2	Galactose
844.3	2.3	Galactose
916.7	3.5	Galactose
1007.7	8.6	Galactose
1248.8	10.3	Galactose

4. Conclusions

We successfully synthesized silver nanoparticles on unpolished silicon substrate using single step microwave-assisted technique. The silver nanoparticles were firmly deposited on the substrate, therefore it is suitable for SERS substrate. The unpolished silicon substrate has micro-size chambers that are useful for Raman measurement. The silver nanoparticles on silicon substrate is around 88 nm in diameter size. Our experiment using corn syrup reveals that our SERS substrate enhance the Raman peaks of glucose and galactose by more than 16 times. Our SERS substrate can be further improved to get high Raman enhancement.

Acknowledgments

This work was done with research grant financial support from Deputy of Engineering Sciences, Indonesian Institute of Science through *Prioritas Nasional* Project financial year 2021. The first author is the main contributor for this work and contributed to all steps of research. The last two authors are supporting contributors that support sample preparation and manuscript preparation.

References

[1] Zhang X-F, Liu Z-G, Shen W and Gurunathan S 2016 *IJMS* **17** 1534
[2] Pilot, Signorini, Durante, Orian, Bhamidipati, and Fabris 2019 *Biosensors* **9** 57
[3] Amendola V, Pilot R, Frasconi M, Maragò O M and Iatì M A 2017 *J. Phys.: Condens. Matter* **29** 203002
[4] Shi R, Liu X and Ying Y 2018 *J. Agric. Food Chem.* **66** 6525–43
[5] Loiseau A, Asila V, Boitel-Aullen G, Lam M, Salmain M and Boujday S 2019 *Biosensors* **9** 78
[6] Hwang J and Yang M 2018 *Sensors* **18** 4076
[7] Novikov S M, Popok V N, Evlyukhin A B, Hanif M, Morgen P, Fiutowski J, Beermann J, Rubahn H-G and Bozhevolnyi S I 2018 *J. Phys.: Conf. Ser.* **1092** 012098
[8] Sofiane S and Mounir S 2019 *Materials Letters* **254** 112–5
[9] Saha S, Ghosh M, Dutta B and Chowdhury J 2016 *J. Raman Spectrosc.* **47** 168–76
[10] Laariedh F, Sow I, Ferchichi A, Zelsmann M and Boussey J 2015 *Microelectronic Engineering* **145** 124–7
[11] Wu W, Liu L, Dai Z, Liu J, Yang S, Zhou L, Xiao X, Jiang C and Roy V A L 2015 *Sci Rep* **5** 10208
[12] Venkatesham M, Ayodhya D, Madhusudhan A, Santoshi Kumari A, Veerabhadram G and Girija Mangatayaru K 2014 *J Clust Sci* **25** 409–22

Fabrication of UV-VIS photodetector based on the ZnO/MoS_2 and ZnO/WS_2 heterostructures

Suci Mufidah Winata[1], **Vivi Fauzia**[1,*]

[1]Department of Physics, Faculty of Mathematics and Natural Science, Universitas Indonesia, Indonesia

*Corresponding author: vivi@sci.ui.ac.id

Abstract. Two-dimensional layers of the transition metal dichalcogenides (2D TMDs) such as molybdenum disulfide (MoS_2) and tungsten disulfide (WS_2) attract much attention due to their potential applications in optoelectronic devices. MoS_2 and WS_2 with tunable bandgap are promising materials to enhance the light absorption range on ZnO-based photodetector. In this work, we report that the exfoliated MoS_2 and WS_2 were deposited on the surface ZnO nanorods for UV-vis photodetector application. The photoelectrical response of photodetector was observed using electrometer under 325, 505, and 625 nm laser diodes as UV and visible light sources. This study shows that the addition of MoS_2 and WS_2 could improve the performance of the ZnO photodetector due to higher photocurrent. The increase in both dark current and photocurrent is not yet clearly understood, but it may be that the edge sites of MoS_2 and WS_2 play a vital role to reduce the natural defects of ZnO; hence the mobility of charge carriers was improved.

1. Introduction

Photodetectors are an essential element of light sensor technology and are widely used in various applications, such as imaging, optical communication, optoelectronic integrated circuits [1], missile detection, fire alarms, and radiation monitoring [2]. Among the many types of photodetectors, such as GaN, Si, SiC and AlGaN, zinc oxide (ZnO) is one of the leading choices for producing high-performance UV photodetectors due to wide bandgap (3.37 eV), considerable binding energy (60 meV), low cost, ease of manufacture, and non-toxic [3]. ZnO is often used as a UV photodetector in various structural configurations, including quantum dots, nanotetrapods, nanotubes, nanowire and nanorods [4]. ZnO one-dimensional (1D) nanostructure arrangement plays an essential role in photodetector performance because of its high surface to volume ratio, high optical gain, high response speed compared to the bulk form of ZnO [4]. In recent years, many ZnO nanostructure-based photodetector studies have been reported. Singh et al. reported a ZnO nanowire-based photodetector grown on nanopillar SiO_2 [5]. Luo et al. reported the configuration of UV photodetector heterojunction based on n-ZnO nanowire/p-silicon [6]. Husham et al. reported a UV detector based on ZnO nanorods grown on seed layers on the p-Si substrate (100)[7]. However, ZnO-based photodetectors still have challenges in high dark current and slow response. To overcome this challenge, metal oxide, organic material and transition metal dichalcogenides are used to improve ZnO-based photodetectors [2].

The heterostructure between ZnO and two-dimensional transition metal dichalcogenides material such as MoS_2 and WS_2 is an exciting option to increase photodetector's performance [8]. MoS_2 and WS_2 show unique properties such as tunable bandgap, spin-orbit coupling, large surface area, and many edge active sites, making it attractive for fundamental study and optoelectronic applications [9,10]. MoS_2 and WS_2 layers have strong covalent bonds on basal planes and weak van der Waals (vdW) bonds between layers; hence they are easily exfoliated through mechanical and liquid exfoliation methods [11] [12]. Moreover, a built-in electric field could be formed in heterojunction to assist photocarrier separation and expand the light absorption range [13][14]. Hsiao et al. have observed the increased photocurrent of ZnO-based photodetectors after being coated by the MoS_2 layer; hence the ratio of photocurrent and dark current was improved [15]. As far as the author's knowledge, there are only a few reports regarding the fabrication of ZnO/MoS_2 and ZnO/WS_2 nanocomposite structures as photodetectors. This heterojunction is expected to improve photodetector performance. In addition, MoS_2 and WS_2 nanosheets were prepared using a low-cost and straightforward liquid exfoliation [7,10].

2. Experimental methods

2.1. Synthesis of ZnO nanorods and MoS_2 and WS_2 exfoliation

Analytical grade zinc acetate dihydrate ($C_4H_6O_4Zn.2H_2O$, ZAD), zinc nitrate tetrahydrate ($N_2O_6Zn.4H_2O$, ZAD), and hexamethylenetetramine ($C_6H_{12}N_4$, HMT) were purchased from Merck Millipore. The synthesis of ZnO nanorods onto glass substrates containing ITO coplanar electrodes (purchased from Ossila) was initiated by dissolving 0.2 M ZAD in deionized water then placing it in an ultrasonic nebulizer (GEA 402A1, 1.7 MHz). The seed solution was sprayed onto the substrate on a hot plate at 450°C for 10 min, and the sample was subsequently annealed at the same temperature for one h. The samples were then immersed in a 10-ml aqueous growth solution containing an equimolar of 0.05 M ZNT and HMT and heated in an oven at 95°C for 6 h. The resulting ZnO nanorods were then washed several times with deionized water and dried.

MoS_2 and WS_2 nanosheets were exfoliated from bulk powder (MoS_2 SKU 234842, 2μm 99%; WS_2 SKU 243639, 2 μm 99%) from Sigma-Aldrich. MoS_2 and WS_2 nanosheets were provided through liquid-phase exfoliation method using an ultrasonic homogenizer (FS-540N 270W). 2 mg/ml MoS_2 and 3 mg/ml WS_2 were mixed in a mixture of deionized water and isopropyl alcohol with a ratio of 4:1 and then placed in a beaker glass in a homemade cooling system with a vibration time of 6 s ON and 4 s OFF for 8 h. Furthermore, MoS_2 and WS_2 nanosheets were separated from the unexfoliated ones via a centrifugation process for 15 min with a speed of 14100 RCF. The yellow-green supernatant was then collected as a final product.

2.2. Fabrication of UV-vis PDs

The photodetector consisted of ZnO/MoS_2 and ZnO/WS_2 heterostructure on ITO interdigitated electrodes, shown in Fig. 1. The heterostructure was fabricated by spin-coated 100μL of exfoliated MoS_2 and WS_2 solutions twice onto ZnO nanorods at 500 rpm and 1000 rpm for 5 s and 15 s annealed on a hotplate at 100°C for 45 min in ambient air.

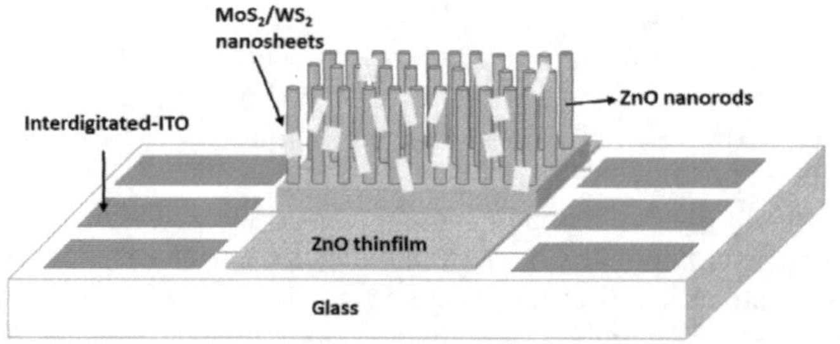

Figure 1. Schematic structure of ZnO/MoS_2 and ZnO/WS_2 -based photodetector.

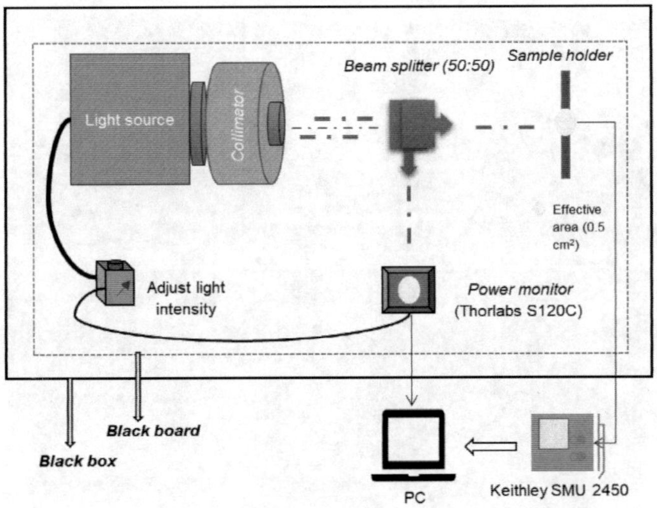

Figure 2. Photodetector's performance measurement set up.

2.3. Characterizations

The morphological properties of the samples were observed using a field emission electron microscope (FESEM; Zeiss Supra 55 V P) and a transmission electron microscope (TEM; FEI 200 kV Tecnai G20 S Twin). The microstructural properties were analyzed using X-ray diffraction (XRD; PANalytical: X'Pert Pro diffractometer with Cu Kα radiation at λ =1.54 Å). Raman spectra were recorded using a WITec Raman microscope with a 532 nm laser source. Optical absorption was recorded using a Thermo Fisher Scientific GENESYS 10S. The photodetector performances were observed by measuring the current-voltage (I-V) relationship using a Keithley 2400 electrometer (Thorlabs) from -2 to +2 V in the black box. The I-V measurement was performed in the dark and under three kinds of laser sources (365, 505 and 625 nm) with an optical power of 0.2 mW on an active area of 0.5 cm^2, as shown in Fig. 2. All characterizations were performed at room temperature.

3. Result and discussion

3.1. Morphological properties

The surface morphologies of ZnO nanorods, ZnO/MoS_2, and ZnO/WS_2 are shown in Figure 3(a)-(c), respectively. FESEM images show that all ZnO nanorods were uniformly grown onto the substrates with high-density hexagonal nanorods in diameters of 109-120 nm. Figures 3 (b) and (c) show the rougher ZnO nanorods after spin-coated by MoS_2 and WS_2 nanosheets. The TEM images in Figures 3(d) and (e) show the 3-4 layers of MoS_2 and WS_2 with the interlayer distance of 0.62 nm and 0.61 nm, corresponding to d-spacing of (002) crystal planes [16].

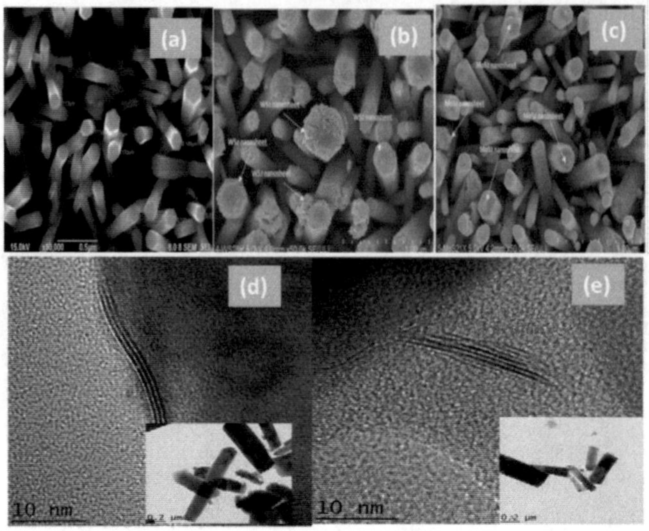

Figure 3. FESEM images of (a) ZnO, (b) ZnO/WS_2, and (c) ZnO/MoS_2 and TEM images of (d) WS_2 and (e) MoS_2.

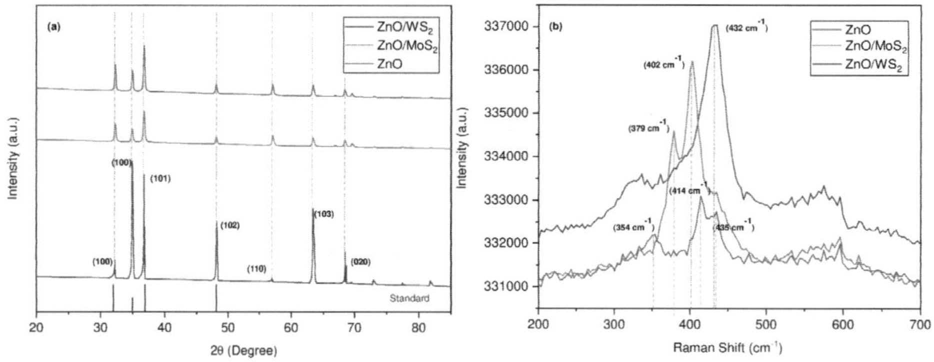

Figure 4. (a) XRD pattern and (b) Raman spectra of ZnO, ZnO/MoS_2, and ZnO/WS_2

3.2. Structural properties

The XRD patterns of pristine ZnO, ZnO/MoS_2 and ZnO/WS_2 heterostructures are shown in Fig. 4(a). Based on the ICSD 98-006-7454 standard, all samples had a single ZnO phase with a hexagonal wurtzite crystal structure. The diffraction peaks at $2\theta = 31.76^o$, 34.42^o, 36.25^o, 47.53^o, 56.59^o, 62.85^o, 66.37^o corresponding to the (100), (002), (101), (102), (110), (103), (020) crystal plane, respectively. The peak of the (002) crystal plane has the highest intensity compared to other diffraction peaks, confirming that the ZnO has a c-axis as preferred orientation. Lattice parameters, unit cell volume, and average crystallite sizes of ZnO were then calculated from the (002) peaks, as shown in Table 1. The addition of MoS_2 and WS_2 does not significantly change the pattern. In addition, no MoS_2 or WS_2 phases were found, which may be due to the MoS_2 and WS_2 nanosheets concentration being too small to be detected by the XRD instrument [17].

Table 1. Microstructure data of ZnO (002) crystal plane.

	2θ	a=b (Å)	c (Å)	Volume ($Å^3$)	d-*spacing*
Standard	34.42	3.25	5.20	47.63	2.60
ZnO	34.95	3.24	5.22	47.64	2.56
ZnO/MoS$_2$	34.93	3.25	5.21	47.79	2.56
ZnO/WS$_2$	34.03	3.25	5.21	47.83	2.56

Fig. 4(b) shows the Raman spectrum of pristine ZnO, ZnO/MoS_2, and ZnO/WS_2. The Raman spectrum of pristine ZnO reveals one high peak at 432 cm^{-1} (E_2) correlated to the ZnO wurtzite structure [18]. The heterostructure of ZnO/MoS_2 and ZnO/WS_2 exhibit two additional peaks, E_{2g}^1 and A_{1g}. For ZnO/MoS_2, the A_{1g} peak at 403 cm^{-1} represents the out-of-plane vibration while the E_{2g}^1 peak at 379 cm^{-1} represents the in-plane vibrational mode[19,20]. ZnO/WS_2, E_{2g}^1 and A_{1g} peaks are located at 414 cm^{-1} and 354 cm^{-1} [21]. Furthermore, the distance between E_{2g}^1 and A_{1g} modes is 24 cm^{-1} and 60 cm^{-1} for ZnO/MoS_2 and ZnO/WS_2, respectively. This result indicates that the few-layer were formed corresponding to TEM images in Fig. 3 [21,22].

3.3. Optical properties

The absorbance spectra of pristine ZnO nanorods and the ZnO/MoS_2 and ZnO/WS_2 heterostructure are shown in Fig. 5(a), while the absorbance spectra of the MoS_2 and WS_2 nanosheets dispersed in solution are shown in Fig. 5(b). The absorbance spectrum for ZnO shows high absorption in the UV light region (300-400 nm) with an absorbance band edge at around 376 nm. The addition of MoS_2 or WS_2 to the surface of ZnO resulted in a widening of the absorption area in the visible region (400-700 nm). This finding is because MoS_2 and WS_2 are low bandgap semiconductors that absorb visible light [23].

The absorbance spectra of the MoS_2 and WS_2 nanosheet solution in Fig. 5(b) show two similar peaks (A and B) corresponding to the optical transition at point K in the Brillouin zone [24]; meanwhile, the C and D peaks in MoS_2 solution correspond to a direct transition of exciton at point M in the Brillouin zone [25]. These absorbance spectra were used to calculate the bandgap with the Tauc equation (1):

$$(\alpha h v)^{\frac{1}{n}} = A\,(hv - E_g) \tag{1}$$

where α is the absorption coefficient, h is Planck constant, v is frequency, E_g is bandgap energy, n is 1/2 for the direct bandgap materials such as ZnO and the exfoliated MoS_2 and WS_2 nanosheets. By extrapolating the linear part of the curve to the horizontal axis as shown as Tauc plots in Fig. 6, the band gaps of MoS_2 and WS_2 nanosheets are 2.39 and 2.35 eV, respectively (see Fig. 6(a) and 6(b)), while the bandgap energy of ZnO, ZnO/MoS_2 dan ZnO/WS_2 are similar in the range of 3.22-3.24 eV (Fig. 6(c)).

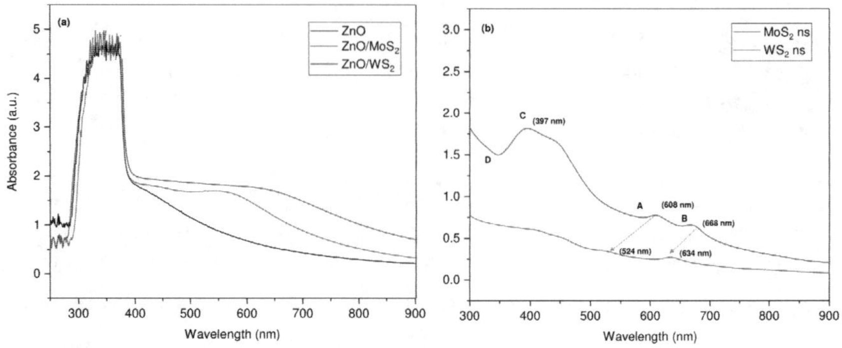

Figure 5. Absorbance spectra of (a) ZnO, ZnO/MoS_2, and ZnO/WS_2 (b) MoS_2 and WS_2 dispersions

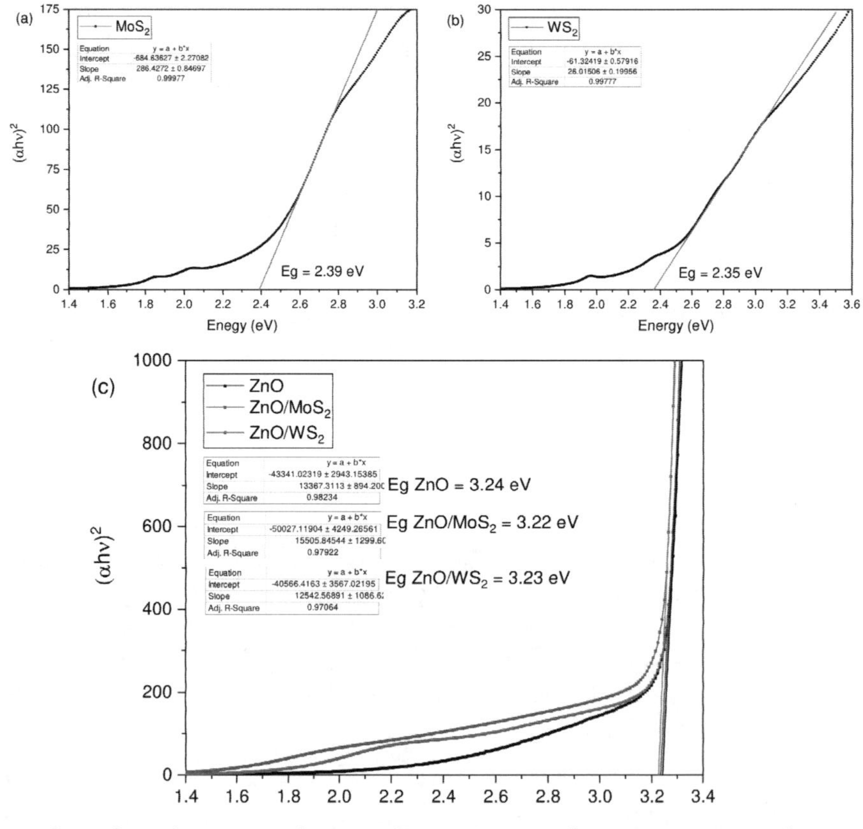

Figure 6. Tauc plot of (a) the MoS_2, (b) WS_2 dispersions and (c) ZnO nanorods, ZnO/MoS_2, and ZnO/WS_2

3.4. Characterization of the fabricated ZnO-based UV-vis detector

Fig. 7 shows the current-voltage (I-V) characteristic of ZnO and ZnO/MoS_2 and ZnO/WS_2 heterostructures in the dark and under light irradiation with the wavelength of 325, 505, and 625 nm. The I-V linear curves indicate the Ohmic contact between ZnO and ITO electrodes [10]. Under bias +2 V, the dark current of pristine ZnO is 3.3 μA and then increased to 7.0 μA when MoS_2 or WS_2 coated ZnO. The dark current is coming from the abundance of natural defects of ZnO, generating an unexpected current [26]. The increase in dark currents is not yet clearly identified but may be due to the surface passivation effect, where the natural defects of ZnO were trapped by the edge sites of MoS_2 and WS_2; hence the mobility of charge carriers were improved [27]. This phenomenon is also observed as the increase of the photocurrent in all wavelengths of the light source. The highest photocurrents were achieved under UV light (325 nm) from ZnO/WS_2 and ZnO/MoS_2 heterostructures of 18.4 μA and 17.3 μA, respectively pristine ZnO (11.8 μA). This highest photocurrent under UV is reasonable due to the highest optical absorbance in the UV region, as shown in Fig. 5(a). High absorbed UV photons were used to excite the electrons from the valence band to the conduction band and left the holes in the valence band. These photogenerated electrons and holes were then transported to the respective electrodes as the photocurrent [26].

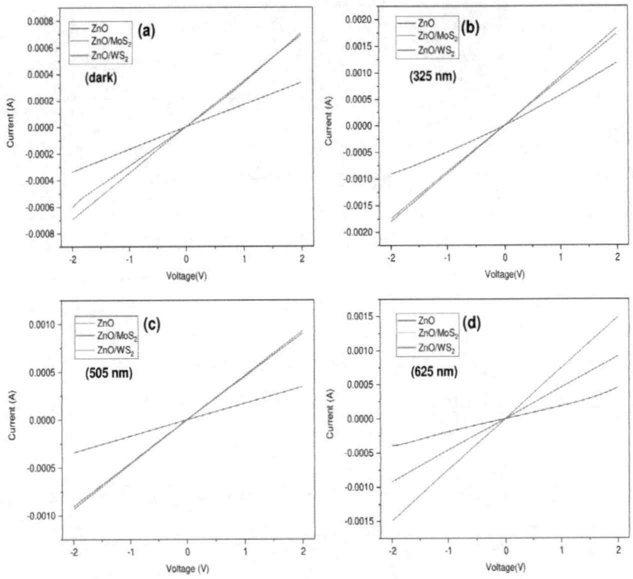

Figure 7. I-V characteristic of ZnO nanorods vs ZnO/MoS_2 and ZnO/WS_2 heterostructure (a) in the dark and under light with a wavelength, (b) 325 nm, (c) 505 nm (d) 625 nm.

The I-V data were used to calculate three photodetector's performance parameters: Responsivity, Detectivity and Sensitivity. Responsivity (R_λ) is defined as the ratio of photocurrent to optical power to the effective area of the photodetector [22].

$$R_\lambda = \frac{\Delta I_{Ph}}{A \times P} \quad (2)$$

where P is the optical power, A is the effective area of the photodetector, I_{dark} is the current in the dark, I_{Ph} is the measured photocurrent, and ΔI_{Ph} is the net photocurrent ($\Delta I_{Ph} = I_{Ph} - I_{dark}$). Detectivity ($D^*$) is defined as the device's ability to detect weak optical signals, given by [22,28].

$$D^* = \frac{R_\lambda \ A^{1/2}}{\sqrt{2 \, e I_{dark}}} \quad (3)$$

where R_λ is responsivity, e is electron charge (1.6×10^{-19} C. D* in units of Jones, which 1 Jones = 1 cmHz$^{1/2}$W^{-1}[9]. In addition, Sensitivity is the ratio of the net photocurrent to the dark current.

$$(S\%) = \frac{I_{light} - I_{dark}}{I_{dark}} \times 100 \quad (4)$$

The increased photocurrent is higher than the increased dark current; hence, the Responsivity, Detectivity and Sensitivity are increased under three kinds of light sources, as shown in Table 2.

Tabel 2. Photodetector performances

λ (nm)	Power (mW)	Sample	Responsivity (A/W)	Detectivity (10^{11} Jones)	Sensitivity (%)
		ZnO	13.1	7,9	205
325	0,2	ZnO/WS_2	18.4	8,6	159
		ZnO/MoS_2	17.3	8,2	151
		ZnO	6.9	4,2	60
505	0,2	ZnO/WS_2	9.1	4,2	28
		ZnO/MoS_2	9.3	4,4	34
		ZnO	4.5	2,7	5
625	0,2	ZnO/WS_2	9.2	4,3	29
		ZnO/MoS_2	14.9	7,1	115

4. Conclusions

In this work, the ZnO nanorods were modified by the exfoliated MoS_2 and WS_2 nanosheets for photodetector application by using the laser sources with a wavelength of 325 nm, 505 nm and 625 nm. MoS_2 and WS_2 nanosheets prepared by a liquid exfoliation method for 8 h consist of 3-4 layers with a bandgap of 2.39 eV and 2.35 eV, respectively. The ZnO/MoS_2 and ZnO/WS_2 heterostructures show improved performance where the Responsivity, Detectivity and Sensitivity were increased compared to the pristine ZnO. The increase in both dark current and photocurrent is not yet clearly understood, but it may be that the edge sites of MoS_2 and WS_2 play a vital role to reduce the natural defects of ZnO; hence the mobility of charge carriers was improved.

Acknowledgments

This work was financially supported by PUTI Q1 NKB-1374/UN2.RST/HKP.05.00/2020 from Universitas Indonesia

References

[1] Zhang C, Wang S, Yang L, Liu Y, Xu T, Ning Z, Zak A, Zhang Z, Tenne R and Chen Q 2012 High-performance photodetectors for visible and near-infrared lights based on individual WS 2 nanotubes *Appl Phys Lett* **100**

[2] Nguyen T T, Patel M, Ban D K and Kim J 2018 Vertically trigonal WS2 layer embedded heterostructure for enhanced ultraviolet–visible photodetector *J Alloys Compd*

[3] Alsultany F H, Hassan Z and Ahmed N M 2016 A high-sensitivity, fast-response, rapid-recovery UV photodetector fabricated based on catalyst-free growth of ZnO nanowire networks on glass substrate *Opt Mater (Amst)* **60** 30–7

[4] Khayatian A, Asgari V, Ramazani A, Akhtarianfar S F, Kashi M A and Safa S 2017 Diameter-controlled synthesis of ZnO nanorods on Fe-doped ZnO seed layer and enhanced photodetection performance *Mater Res Bull* **94** 77–84

[5] Singh S and Park S-H H 2017 Fabrication and properties of ZnO nanorods based MSM UV detectors on silicon substrates *Optik (Stuttg)* **137** 96–100

[6] Luo L, Zhang Y, Mao S S and Lin L 2006 Fabrication and characterization of ZnO nanowires based UV photodiodes *Sensors Actuators, A Phys* **127** 201–6

[7] Husham M, Hamidon M N, Paiman S, Abuelsamen A A, Farhat O F and Al-Dulaimi A A 2017 Synthesis of ZnO nanorods by microwave-assisted chemical-bath deposition for highly sensitive self-powered UV detection application *Sensors Actuators A Phys* **263** 166–73

[8] Ning L, Jiang T, Shao Z, Ding K, Zhang X and Jie J 2018 Light-trapping enhanced ZnO-MoS2core-shell nanopillar arrays for broadband ultraviolet-visible-near infrared photodetection *J Mater Chem C* **6** 7077–84

[9] Manzeli S, Ovchinnikov D, Pasquier D, Yazyev O V and Kis A 2017 2D transition metal

dichalcogenides *Nat Rev Mater* **2** 17033

[10] Zhang Y, Xu J, Shi S, Gao Y, Zhao X, Wei C, Zhang X and Li L 2017 Photoelectric response properties under UV/red light irradiation of ZnO nanorod arrays coated with vertically aligned MoS2 nanosheets. *Nanotechnology* **28** 415202

[11] Fan X, Xu P, Li Y C, Zhou D, Sun Y, Nguyen M A T, Terrones M and Mallouk T E 2016 Controlled Exfoliation of MoS2 Crystals into Trilayer Nanosheets *J Am Chem Soc* **138** 5143–9

[12] Duan X, Wang C, Pan A, Yu R and Duan X 2015 Two-dimensional transition metal dichalcogenides as atomically thin semiconductors: Opportunities and challenges *Chem Soc Rev* **44** 8859–76

[13] Mukherjee S, Maiti R, Midya A, Das S and Ray S K 2015 Tunable Direct Bandgap Optical Transitions in MoS2 Nanocrystals for Photonic Devices *ACS Photonics* **2** 760–8

[14] Duong D L, Yun S J and Lee Y H 2017 Van der Waals Layered Materials: Opportunities and Challenges *ACS Nano* **11** 11803–30

[15] Hsiao Y J, Fang T H, Ji L W and Yang B Y 2015 Red-Shift Effect and Sensitive Responsivity of MoS2/ZnO Flexible Photodetectors *Nanoscale Res Lett* **10** 1–5

[16] Liu W, Benson J, Dawson C, Strudwick A, Raju A P A, Han Y, Li M and Papakonstantinou P 2017 The effects of exfoliation, organic solvents and anodic activation on the catalytic hydrogen evolution reaction of tungsten disulfide *Nanoscale* **9** 13515–26

[17] Renaud G, Lazzari R and Leroy F 2009 Probing surface and interface morphology with Grazing Incidence Small Angle X-Ray Scattering *Surf Sci Rep* **64** 255–380

[18] Gao Q, Dai Y, Li C, Yang L, Li X and Cui C 2016 Correlation between oxygen vacancies and dopant concentration in Mn-doped ZnO nanoparticles synthesized by co-precipitation technique *J Alloys Compd* **684** 669–76

[19] Krishnan U, Kaur M, Singh K, Kumar M and Kumar A 2019 A synoptic review of MoS 2 : Synthesis to applications *Superlattices Microstruct* **128** 274–97

[20] Li H, Zhang Q, Yap C C R, Tay B K, Edwin T H T, Olivier A and Baillargeat D 2012 From bulk to monolayer MoS 2 : Evolution of Raman scattering *Adv Funct Mater* **22** 1385–90

[21] Butanovs E, Vlassov S, Kuzmin A, Piskunov S, Butikova J and Polyakov B 2018 Fast-Response Single-Nanowire Photodetector Based on ZnO/WS2Core/Shell Heterostructures *ACS Appl Mater Interfaces* **10** 13869–76

[22] Nazir G, Khan M F, Akhtar I, Akbar K, Gautam P, Noh H, Seo Y, Chun S H and Eom J 2017 Enhanced photoresponse of ZnO quantum dot-decorated MoS2thin films *RSC Adv* **7** 16890–900

[23] Terrones M, Voshell A and Rana M M 2018 Review of optical properties of two-dimensional transition metal dichalcogenides *Wide Bandgap Power and Energy Devices and Applications III* ed M Matin, A K Dutta and S Chowdhury (SPIE) p 21

[24] Zhu B, Chen X and Cui X 2015 Exciton binding energy of monolayer WS2 *Sci Rep* **5**

[25] Rahmanian E and Malekfar R 2017 Size-dependent optical response of few-layered WS 2 nanosheets produced by liquid phase exfoliation *Eur Phys J Appl Phys* **77** 30401

[26] Zheng M, Gui P, Wang X, Zhang G, Wan J, Zhang H, Fang G, Wu H, Lin Q and Liu C 2019 ZnO ultraviolet photodetectors with an extremely high detectivity and short response time *Appl Surf Sci* **481** 437–42

[27] Butanovs E, Vlassov S, Kuzmin A, Piskunov S, Butikova J and Polyakov B 2018 Fast-Response Single-Nanowire Photodetector Based on ZnO/WS2 Core/Shell Heterostructures *ACS Appl Mater Interfaces* **10** 13869–76

[28] Datskos P G and Lavrik N V. 2003 Detectors - Figures of Merit *Powder Technol* **177** 170–84

Journal of Physics: Conference Series **2274** (2022) 012005 doi:10.1088/1742-6596/2274/1/012005

Phase detection autofocus (PDAF) image sensor as applied to 3D imaging

Suganda Jutamulia[1,*]

[1]Omnivision Technologies, Inc., USA

[*]Corresponding author: suganda@sbcglobal.net

Abstract. Auto-focus in a camera can be performed by comparing the image formed by light passing through a half of the imaging lens to the image formed by light passing through the other half of the imaging lens. A PDAF image sensor has an array of pixels. A microlens covers a pixel consisting of two photodiodes. In a camera using a PDAF image sensor, a photodiode receives light passing through a half of the imaging lens and the other photodiode receives light passing through the other half of the imaging lens. Thus, comparing light received by two photodiodes, a PDAF image sensor is capable of performing auto-focus of the camera. The PDAF image sensor is also capable of performing 3-D imaging.

1. Introduction

There are many solutions for capturing 3D images including stereo camera, time-of-flight, and structure light. In the solution using structure light, the structure light emitted by vertical cavity surface emitting lasers (VCSELs) is projected on a target surface. In the solution using time-of-flight, a laser pulse is projected on the target surface. If the target is a human face, the projected laser light may cause injury to the human face.

3D imaging of human face recently is used for 3D face recognition in unlocking a smart phone. The stereo camera solution needs a sufficient parallax, which in turn needs a relatively large base line between two cameras. The time-of-flight solution needs an advanced electronic processor to determine the extremely short time difference between the light firing time and the reflected light receiving time. The structure light solution needs a more complex optical system for projecting the structure light, thus it may be relatively large for being use in smart phones.

New solutions for capturing 3D images, especially for being used in smart phones, are proposed in a US patent [1] and discussed in this paper. The new solutions may not change the size of the device such as a smart phone, and ideally may not add extra hardware elements to complicate the device.

2. Results and Discussions

A digital camera may include a lens and an image sensor. The lens formed the image of an object on the image sensor. The digital camera may be a camera installed in a smart phone, or may be a digital SLR (single lens reflex) camera. A phase detection autofocus (PDAF) image sensor may be used in the camera [2]. Together with a lens driven by an actuator, the PDAF image sensor may perform auto focusing. The inherent functions of the PDAF image sensor may be used for 3D imaging. In this manner, no extra hardware elements are required for performing 3D imaging.

Figure 1 illustrates the principle of PDAF. A lens images an object on an image plane. Left light passes the left half of the lens. Right light passes the right half of the lens. If a left mask is placed to block left light, only right light will form an image on the image plane. If a right mask is placed to block right light, only left light will form an image on the image plane.

Figure 1 shows the lens at an out-of-focus position. Focus is formed at the front of the image plane. On the image plane, a first image in the left image formed by left light is shifted in horizontal direction from a second image in the right image formed by right light. When the lens is at an in-focus position, focus is formed at the image plane. On the image plane, the first image in the left image formed by left light overlaps with the second image in the right image formed by right light.

Figure 2 illustrates a PDAF pixel of a PDAF CMOS image sensor [3,4]. The pixel has two isolated left and right photodiodes and a microlens. The incident light comprises right light and left light as shown in Fig. 1. The micro lens focuses incident right light to the right photodiode of the pixel. The same micro lens focuses incident left light to the left photodiode of the pixel.

When the right image is shifted from the left image as shown in Fig. 1, the value of the right photodiode is different from the value of the left photodiode. When the right image of Fig. 1 overlaps with the left image of Fig. 1, the value of the right photodiode is the same as the value of the left photodiode. Therefore portions of an object at an in-focus plane of Fig. 3 may be identified by comparing the values of the right photodiode and the left photodiode of the PDAF pixel. If they are the same, the pixel of the PDAF pixel array detects the image of the portion of the object at the in-focus plane as shown in Fig. 4. By moving the in-focus plane step by step, a 3D image of an object can be constructed.

3. Conclusions

This paper discusses the proposal of using a smart phone equipped with PDAF image sensor to capture a 3D image, without any additional hardware or optical system to the smart phone.

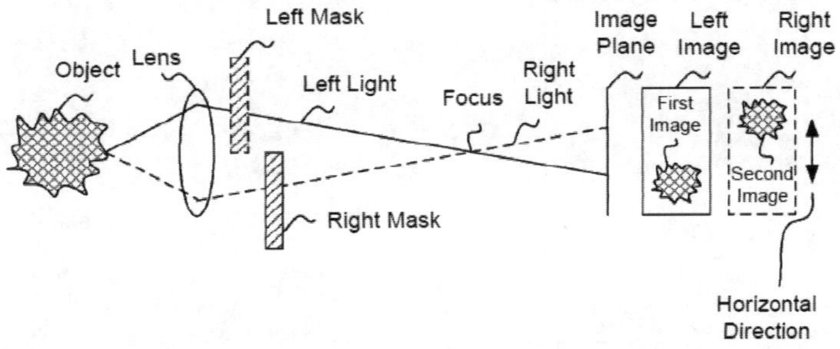

Figure 1. Principle of PDAF.

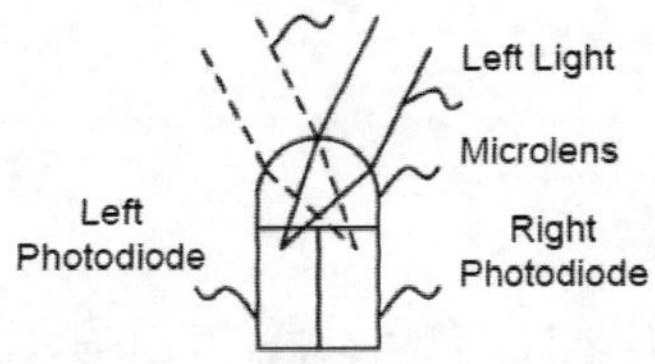

Figure 2. PDAF pixel of PDAF image sensor.

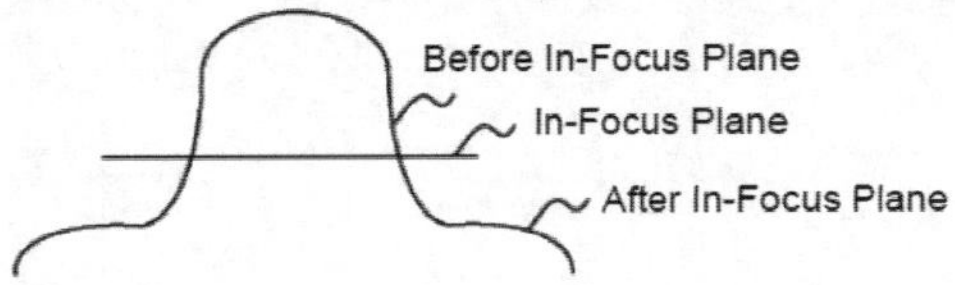

Figure 3. Portions of object at in-focus plane.

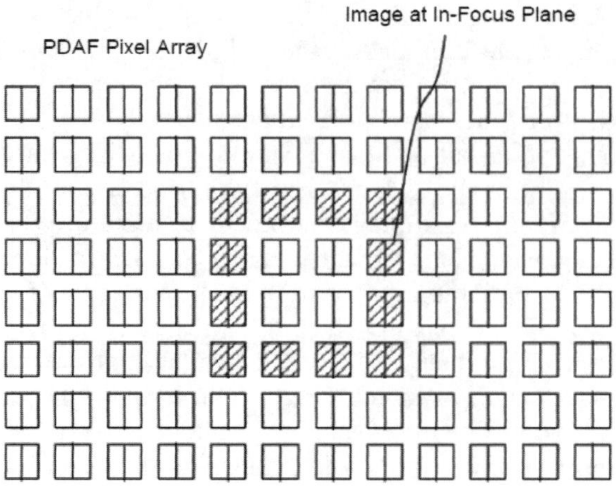

Figure 4. PDAF pixel array of PDAF image sensor.

References

[1] Chan A, Liu L, and Jutamulia S 2021 US Patent 10,996,426
[2] Iizuka K 2019 *Engineering Optics* (Basingstoke: Springer Nature)
[3] Jang J, Yoo Y, Kim J, and Paik J 2015 *Sensors* **15** 5747
[4] Chataignier G, Vandame B, and Vaillant J 2019 *Optics Express* **21** 30486

The improvement design of two-colour pyrometry for pulsed laser deposition system

T. Liang[1], K. Ningsih[1,2a], J. I. Mackenzie[1], J. Prentice[1a], M. Yasin[2b]

[1] Optoelectronics Research Centre, The University of Southampton, Highfield Campus, Southampton SO17 1BJ, UK
[2] Department of Physics, Faculty of Science and Technology, Airlangga University, 60115, Indonesia

Corresponding author: kn1n17@soton.ac.uk[a], yasin@fst.unair.ac.id[b]

Abstract. In this work, we develop a digital pyrometry with an automatic control system. A two-colour pyrometer is applied to measure the substrate surface temperature in the pulsed laser deposition (PLD) system. A digital computer and two-colour photodetector are coupled to obtain the substrate surface temperature and the output signal from the respective detectors with high accuracy. These sets enable us to determine the substrate temperature during the process of deposition remotely and potentially allow us to develop control routines for maintaining a constant temperature throughout the process. The designed system has a broad spectral response range from 0.32 to 1.7μm and achieves a deficient full-scale temperature error of <1.8^{0}C. Factors affecting the temperature measurements are studied to identify the sensor limitations, such as possible damage on the two-colour photodetectors, spectral losses, responsivity, and the distance between the photodetector end to the substrate. Finally, this pyrometric system is applied in the PLD process to maintain a stable substrate surface temperature.

1. Introduction

Thin film technology has played a vital role in superconducting optoelectronics research. Many inventions and methods have been developed in this area: vacuum evaporation deposition, ion beam sputtering, magnetron sputtering deposition, metal-organic chemical vapour deposition, and sol-gel method [1]. However, these methods still have limitations in making thin films, especially in terms of efficiency and therefore cannot satisfy the development of various thin-film research. With the development of the high-power laser, the pulsed laser deposition (PLD) methods emerged. This method gives promising results in the field of optoelectronics, especially in the fabrication of thin film superconductors and the the fabrication of active gain elements as amplifiers and laser oscillators [2].

In the pulsed laser deposition process, a pulsed laser beam focuses on ablating the target and generating on the target plasma. The high energetic plasma plume creates a plume that travels in a vacuum toward a heated substrate then the deposited material grows into a crystalline film [3]. However, the high energetic plasma plume causes surface damage and raises the film defect potential when it is not under the correct condition. Therefore, a pyrometry system needs to be integrated into the PLD process for thermal equilibrium, especially the stable substrate surface temperature. Thermal stability should be achieved for steady and sustainable film growth to produce a high-quality thin film [4].

Pyrometer is a remote sensing thermometer to measure the surface temperature of an object based on the amount of heat radiated [5]. A conventional pyrometer measures the surface temperature based on a single wavelength and surface emissivity. Nevertheless, it is not easy to obtain the emissivity of an unknown surface due to some variables affecting the data accuracy in different conditions. Meanwhile, the two-colour pyrometer can measure the actual temperature regardless of the target's emissivity, which the measurement result will not be changed [6-8].

The first process of the PLD in this experiment is focusing the pulsed laser on the surface of the target. The target absorbs short pulses and rapidly raises the temperature until the target evaporates to form a high-temperature, high-density plasma. The next step is about the moving of plasma [9-12]. Further ionization of the plasma continues to increase its temperature and density while the volume begins to expand. At the last step, the plasma nucleates and deposits on the substrate after bombarding the substrate to form a uniform thin film [9,10]. However, the bombardment of high-energy particles can cause damage to the substrate surface and increase the uniformity while growing the thin films. [11,12]. Many factors determine the deposition of thin films such as the temperature of the substrate as one of the most critical factors [12]. Therefore, a pyrometric system that can maintain a stable substrate surface temperature in the PLD process needs to be developed.

2. Experiment Set up

This project focuses on designing the pyrometry system for the PLD process. It has two goals: one is to establish a theoretical deduction process for the two-colour method, and the other is to improve the measurement accuracy of the existing setup. Overall, this method combines quantitative analysis of thermal radiation theory and two-colour temperature measurement to design an automatic temperature-controlled system for the PLD process. Two kinds of materials, Sapphire and Yttrium Aluminum Garnet (YAG), were used in this work which are the equal in size with dimensions of10mm × 10mm × 10mm, were used as substrates in this work. Sapphire is used as a substrate because it has a melting point of up to 2040 degrees Celsius and is chemically inert at high temperatures. In addition, this material also has a high level of hardness and is transparent in the visible spectrum. YAG is used as a substrate because it is an important inorganic non-metallic functional material and is a high-quality matrix material of solid-state lasers. Hence, these two substrate materials are quite representative for research based on the temperature measurement system like in this work. [13, 14].

In this work, the design of the temperature measurement system consists of a computer integrated with power meters, an oscilloscope, a two-colour detector, and a simple optical system. The two-colour pyrometer (Hamamatsu K1713-09) was used to measure the temperature of the substrate remotely. The oscilloscope draws the graph of the signal voltage over time, which is often used to display the electronic signal. The thermal radiation emitted from the substrate was detected by a pyrometer. The two photodiodes, Si and InGaAs, of the pyrometer generate two output voltages. The output voltage signals of the two photodiodes were displayed on the oscilloscope.

Journal of Physics: Conference Series **2274** (2022) 012006 doi:10.1088/1742-6596/2274/1/012006

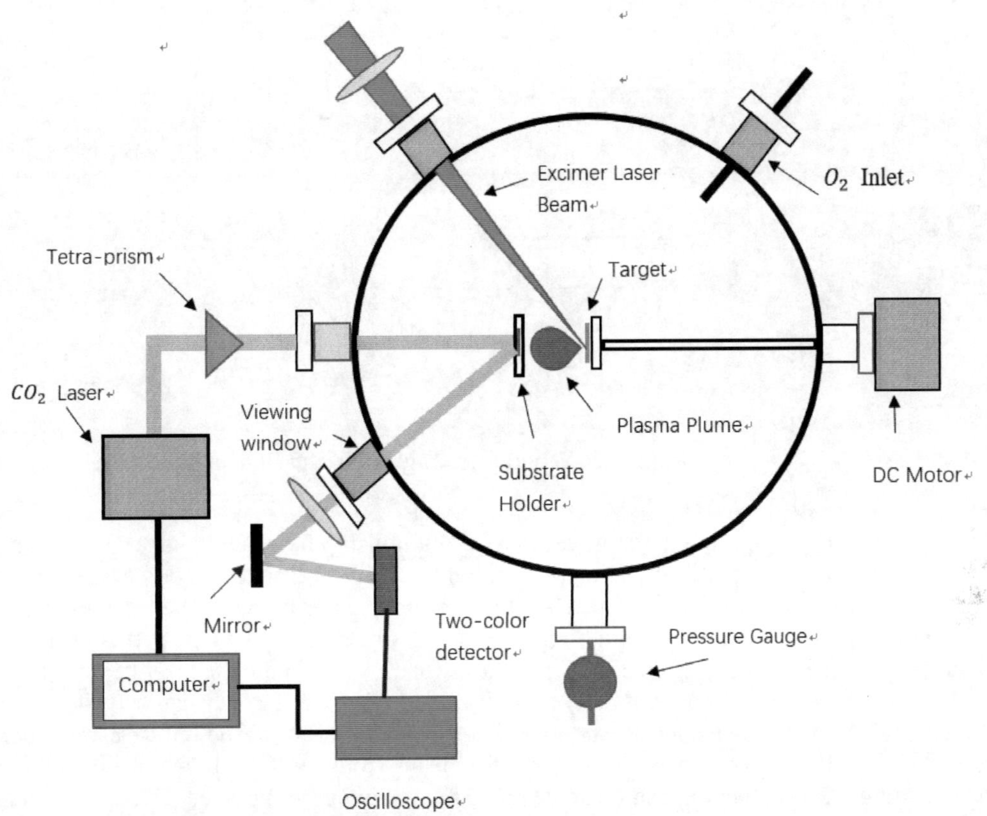

Figure 1. The schematic of the pyrometry system for the PLD process.

For the PLD process shown in Figure.1, a KrF (Krypton Fluoride) excimer laser (Coherent–COMPex PRO 110) with 20 ns pulsewidth and maximum output energy of 750 mJ operating at 248 nm was used to ablate the target materials [11]. A continuous-wave (CW) CO2 laser (Rofin OEM10iX) was used to heat the substrate in a large temperature range (exceeding 1000 °C). The CW laser assembly consists of three components, a laser tube, a fine-tuned optical correction system, and an integrated power unit. Even though this laser has a very high and stable energy output (maximum 1.3 kW) that can heat the substrate uniformly, the measured output energy used in this experiment is 1 kW. This experiment were carried out in an oxygen environment where the pressure inside the chamber was maintained at 0.02 mbar. The emitted light from the heated substrate is detected by the two-colour detector through the view window.

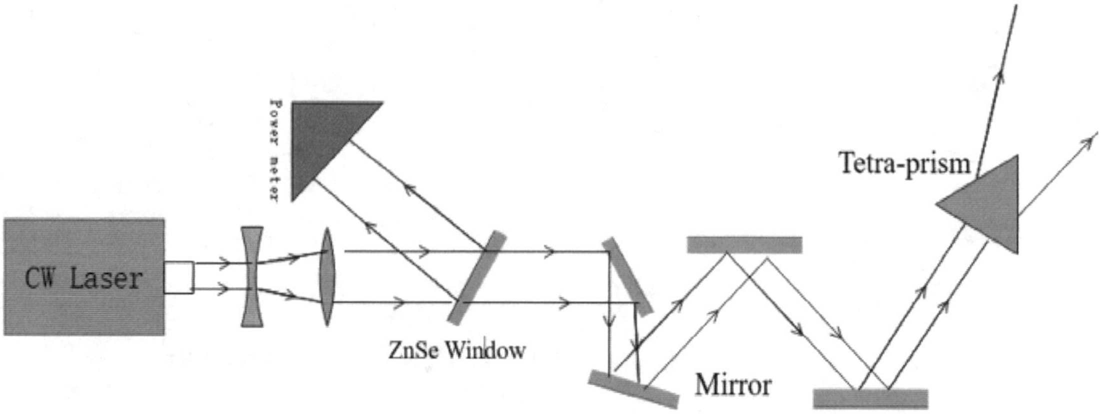

Figure 2. Schematic diagram of the light path.

The first step of the experiment is to quantify the relationship between the energy of each part of the PLD, so it needs to measure the power at two different positions through the power meters, which were integrated into the computational system. The optical path diagram between the laser and the tetra-prism is shown in Fig. 2. In this optical system, there are mainly three parts. The first part is a telescope system consisting of a convex lens and a concave lens, mainly used to collimate the laser beam. The second part is a ZnSe (Zinc Selenide) window, widely used in the high-power laser because of its low absorption coefficient and strong thermal shock resistance. Then the beam touched the tetra-prism by the reflection of four mirrors. The tetra-prism can distribute the beam from the CW laser on the substrate so that uniform heating uniform heating can be achieved. The two integrated power meters used in this experiment have different functions. One power meter is used to measure the reflected power of CW laser by the ZnSe window, and another power meter is used to measure the radiated power of the substrate. These two data were recorded as pick-off and chamber power, respectively. The various ratio between two voltages and the power of the PLD are recorded to build a mathematical model.

3. Mathematical Model

Reliable data support is required to develop mathematical model for quantitative analysis. In this work, the data source was divided into two parts. Primary data is the raw data obtained through experiments. The data was obtained was obtained by changing the laser power and recording the photodiode voltage changes at different laser powers [15]. Secondary data were collected from the journal literature, mainly for the characteristics of the two substrate materials, such as the change of silicon transmittance at different wavelengths. It also includes the basic characteristics of the two photodiodes of the two-colour detector [16, 17]. These data summaries were used to develop a mathematical model of the temperature measurement system. According to the theory of the blackbody spectral radiation, with the increase of the temperature, the maximum of the spectral specific radiation moves to the shorter wavelength. Moreover, the peak of wavelength $\boldsymbol{\lambda_{max}}$ given by the Wien's law shown as below:

$$\lambda_{max} = \frac{b}{T} \qquad (1)$$

where b is the Wien's displacement constant and approximately equal to $2.8978 \times 10^{-3} mk$, λ_{max} is wavelength in meters, and T is the temperature in kelvin.

The other successful formula is based on the Rayleigh-Jeans formula derived from classical electrodynamics and statistical mechanics. In the high-frequency part, the blackbody radiation tends to be infinite, which contradicts the experiment condition. It was called *ultraviolet catastrophe*. The energy density u_v is shown as below formula:

$$u_v = \frac{8\pi v^2 kT}{c^2} \tag{2}$$

where the v is the frequency, k is Boltzmann's constant and T is the absolute temperature of the blackbody. Moreover, to measure the total energy emitted per unit of the surface area from blackbody per unit time maintained at a temperature T, it needs to describe the relationship between the temperature and emitted energy. This is called the *Stefan–Boltzmann law:*

$$\frac{P}{A} = \sigma T^4 \tag{3}$$

where σ is Stefan–Boltzmann constant, $\sigma \approx 5.67 \times 10^{-8}$ W/(m^{2}K^4). The emitted energy is proportional to the fourth power of the absolute temperature of the black body, which can be used to roughly estimate the surface temperature of objects. For the design of pyrometry, this formula is allowed to measure the ideal substrate surface temperature, and it can also be used as a control group to compare the real data.

The last one is also the most important theory for the project is the Planck's law. Focusing on Planck's law, B_v describes the amount of energy emitted at different radiation frequencies. It is the power emitted per unit area of the body, per unit solid angle of emission, per unit frequency. Planck showed that the spectral radiance of a body for frequency v at absolute temperature T is given by:

$$B_v(V,T) = \frac{2hV^3}{c^2}\frac{1}{e^{\frac{hv}{k_BT}}-1} \tag{4}$$

where h is the Planck's constant, c is the speed of light and k_B is the Boltzmann's constant. This equation is also can be expressed by the wavelength λ, and the formula is shown below:

$$B_\lambda(\lambda,T) = \frac{2hc^2}{\lambda^5}\frac{1}{e^{\frac{hc}{\lambda k_BT}}-1} \tag{5}$$

showing the relationship between the temperature T and the wavelength λ. This relationship is the theoretical basis of the following temperature measurement system. The above five formulae are quantitatively and qualitatively described the thermal radiation of blackbody. However, for the real case, it is more commonly used as a graybody [17, 18]. The two-colour detector uses the ratio of the radiant energy of the two bands to determine the object's temperature, and the spectral ratio has a proportional relationship with the substrate surface temperature.

Furthermore, Planck's Law explains the principle that the energy radiated from the surface of a substrate object, assumed as a graybody, is proportional to the temperature that can be expressed in the following formula [17, 18, 19]:

$$B_\lambda(\lambda,T) = \frac{C_1}{\lambda^5}\frac{1}{e^{\frac{C_2}{\lambda T}}-1} \tag{6}$$

The above formula describes the relationship between the amount of spectral radiation at any wavelength and the temperature of the object. If C_1equal to $2hc^2$, then C_2 will be equal to hc/k. The c is the speed of light in which equal to $3 \times 10^8 m/s$, h is the Planck's constant equal to $6.627 \times 10^{-34} m^2 kg/s$. Besides, k is Boltzmann's constant. Hence, the number of C_1 and C_2 are $1.191 \times 10^8 WSr^{-1}\mu m^4 m^{-2}$ and 1.439×10^4 μmK respectively.

Because of the Wien's law shown in the equation (1), the maximum value of $\lambda_{max}T = 2897.8\ \mu mK$, and then the Wien's approximation can be used for the equation (6) and the results below 1%, the function can be given by:

$$\left(e^{\frac{C_2}{\lambda T}} - 1\right)^{-1} = e^{-\frac{C_2}{\lambda T}} \tag{7}$$

For the two-colour pyrometer, the detected signal can be derived from the equation 6 and 7:

$$I_\lambda(\lambda,T) = \frac{K\varepsilon_1 C_1}{\lambda^5} e^{-\frac{C_2}{\lambda T}} \tag{8}$$

where $\boldsymbol{\varepsilon_1}$ is the object's emissivity and depends on wavelengths, K is a constant affected by the detector's optical properties and structural factors. By assuming the substrate as a graybody, $\boldsymbol{\varepsilon_1} = \boldsymbol{\varepsilon_2}$, the surface emissivity on the temperature calculation can be neglected [19,20].

As mentioned before, the voltage of the two photodiodes has been recorded during the experiment. Based on the Ohm's law, where the ratio of two voltages equals the ratio of currents: $\frac{V_2}{V_1} = \frac{I_2}{I_1}$, then, the photodiodes output ratio relationship:

$$R = \frac{V_2}{V_1} = \frac{I_2}{I_1} = \frac{\int_{1.1}^{1.7} I_{\lambda 1}(\lambda,T)_{InGaAs} * S_{InGaAs} * T_{Si} d\lambda}{\int_{0.3}^{1.1} I_{\lambda 2}(\lambda,T)_{Si} * S_{Si} * d\lambda} \tag{9}$$

where V_2 and V_1 are voltages of InGaAs photodiodes and silicon photodiodes, respectively. Each of the photodiodes has a different spectral response range. The InGaAs (S_{InGaAs}) has a spectral sensitivity from 1.1 to 1.7 μm, while the silicon's (S_{Si}) response range is between 0.3 and 1.1 μm. T_{Si} is the transmission of silicon photodiodes at the chosen wavelength. The above equation illustrates that the ratio of currents is proportional to the detected radiation in two respective photodiodes [21].

Compared to the traditional two-colour method, this temperature measurement method relies on the accuracy of the spectral band ratio and the relation between the known temperature and this ratio. For the real case, equation (9) shows the relationship between the detected ratio R and the measured temperature T_R as following

$$T_R = \frac{B}{\ln R - A} + C \tag{10}$$

where R is the photodiodes output ratio described in equation (9). A, B, and C are constants, and the correspondence between them is determined by calibrating the ratio with a known temperature. Since the method assumes that the substrate is a grey body, the calculation process is independent of the emissivity, and only one material needs to be calibrated. The temperature data of the diode can be sent back to the laser power control by the computer to stabilize the temperature and automatically perform a constant, gradual heating process [22, 23].

4. Result and Discussion

4.1. Experimental data analysis

The linear relationship between the measured pick-off power and the chamber power is shown in Fig. 3. This relationship is used to calculate the chamber power in real-time when only the pick-off power is known. Besides, it bridges the mathematical connections between the various parts.

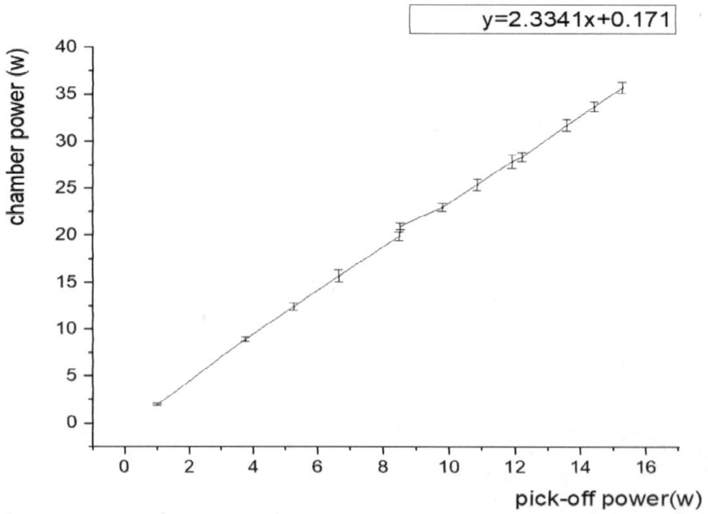

Figure 3. The relationship between pick-off power and the chamber power.

After setting the PLD system, the laser power is gradually increased to the set value. The pick-off power and voltage generated by the two photodiodes were recorded at different set value. The chamber power increased from 8.6 watts to 35 watts when the set value was adjusted from 9 to 24. The substrate temperature was determined from the ratio of the two recorded voltages.

When the surface temperature increases due to the continuous heating of the laser, the spectral band ratio becomes small and presents an inverse relationship. It causes the spectral band ratio to decrease as the temperature increases. Hence, it can be assumed that the error in the temperature data comes from the fluctuation of the spectral band ratio and some noise interference on the oscilloscope's voltage reading, especially for low voltage values.

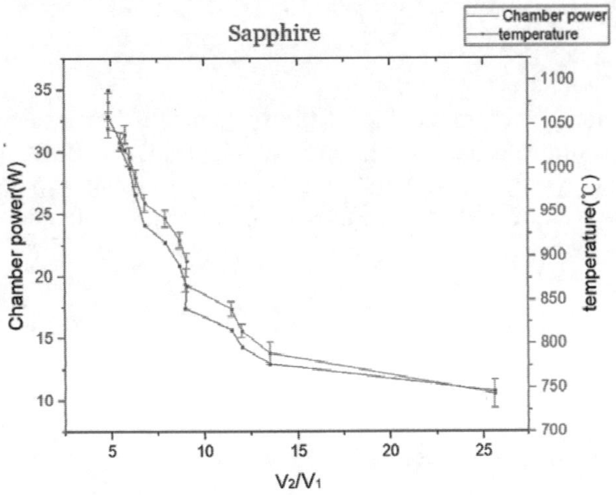

(a)

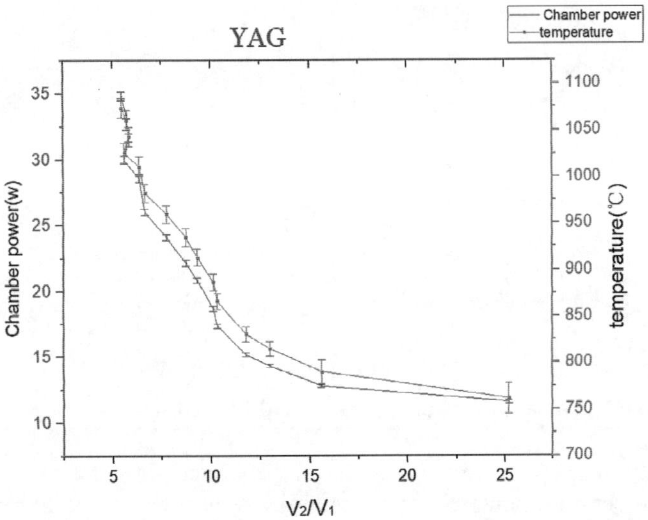

(b)

Figure. 4 The ratio of voltage as a function of substrate temperature and chamber power. (a) for sapphire substrate and (b) for YAG substrate.

After obtaining the relationship between the substrate surface temperature and the voltage ratio, we can determine the voltage and chamber power relationship. From Fig. 4, the power and temperature curves show a similar trend. The Stefan–Boltzmann law stated that the chamber power is proportional to the fourth power of absolute temperature for a blackbody. This relationship is affected by material emissivity and thermal loss for the real system. From the graphs of the two substrate materials, the thermal radiation calculation model gives them a similar curve, and the material properties determine some differences.

In addition, it can be seen from the above figures that the temperature range detected by this system is limited. The minimum detection temperature values for Sapphire and YAG susbtrates are 740 °C and 750 °C, respectively, while the error reached 7% when approching this lowest value range. Meanwhile, for substrate materials in general, a temperature around 600 to 800 °C is required. Therefore, to achieve a better result, the temperature range of the entire automatic temperature control system should be set between 400 and 1200 °C.

4.2 Pyrometry System Test

In order to test the accuracy and practicality of the system, we designed two experiments for further research. In this test, sapphire was used as the substrate material. The first experiment was conducted to check the accuracy of the pyrometric system in measuring the surface temperature of the substrate. The measured temperature was compared with the measurement results obtained using an infrared thermometer and is shown in Figure 5.

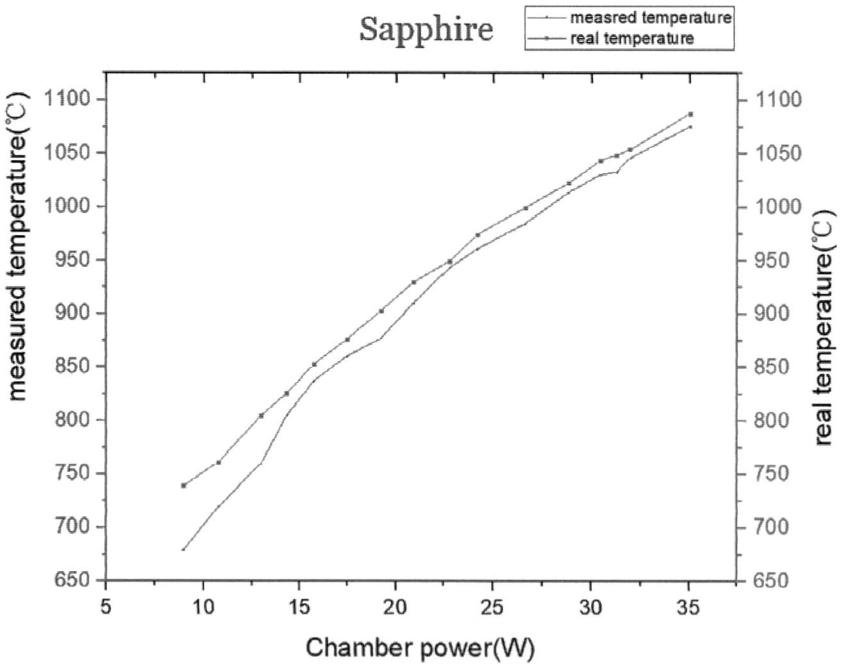

Figure 5. The comparison between real temperature and measured temperature

In Fig. 5, the black line indicates the temperature obtained from the pyrometry system, and the red line indicates the temperature obtained from the measurement using the infrared thermometer. The accuracy of the infrared thermometer is about $\pm 2^0$C. By comparing those two measurement results, we can estimate the accuracy of the proposed pyrometry system. As can be seen in Fig 5, the accuracy of the measured temperature is low at low chamber power region (7-20W). This is related to the oscilloscope noise, where the signal-noise ratio (SNR) is relatively low for the low chamber power region, resulting in a larger temperature error. However, this is not the case in a high chamber power range. The accuracy of the measured remperature can be improved by amplifying the signal through the lock-in amplifier that is placed in front of the two-colour detector.

The second experiment was carried out to check the uniformity of heating of the substrate. This is done by moving the position of the two-colour detector relative to the substrate. By measuring the temperature at nine different positions (3×3 array) on the substrate area, we found that , the temperature difference between different locations is less than 2%.

5. Conclusions

The design of an automatic control pyrometry system for the deposition process of the PLD system to reach a stable substrate surface temperature that can maximize the deposition efficiency of the target has been proposed. We use four different parts to describe the pyrometry system for the PLD process generally. However, this system still has many limitations. First, the project focuses on the temperature measurement principle of the two-colour method and the experimental design of the PLD system. So this research is more at the theoretical and modelling level. As a result, the system error is large, and the temperature measuring range is relatively small (700-1100 °C). Many substrate materials require an optimum deposition temperature of fewer than 700 °C, so the application range of the system is limited. Second, this project lacks an in-depth analysis of the experimental error. There is no timely solution based on the error analysis, which leads to some work to be completed in the future.

For future work we need to improve the experimental setup by adding a lock-in amplifier in front of the two-colour detector, which can extract the signal and amplify the signal up to 1 million times. By amplifying the signal the measuring range of the pyrometry system can be increased and the lower temperature (200-700 °C) can be measured. The second thing to do is to use more substrate material to verify the accuracy of the system and record a large amount of data to support the system. More data support can improve the reliability of the proposed model. Another thing to do is to explore the effects of different material properties on the system as different materials will give different measurement results.

References

[1] Grivas C, May-Smith T C, Shepherd D P, Pollnau M, and Jelinek M, 2004 Broadband single transverse-mode fluorescence sources based on ribs fabricated in pulsed laser deposited Ti: sapphire waveguides (Springer, Journal of Applied Physics A, 79(4-6)) p 1195-1198

[2] Mackenzie J I, Kurilchik S V, Prentice J J, Jacob G J. Carpenter J, Gates J, Smith P.G.R., Riris H., Yu A. W., and Eason RW, 2019 *1.6-µm Er: YGG waveguide amplifiers*, (Proc. SPIE 10896, Solid State Lasers XXVIII: Technology and Devices, 1089604)

[3] Eason, Robert, ed. 2007 Pulsed laser deposition of thin films: applications-led growth of functional materials. John Wiley & Sons

[4] Prentice J J, Grant-Jacob J, Mackenzie J I, Eason R.W, 2018. Yb-Doped Mixed-sesquioxide Films Grown by Pulsed Laser Deposition (Journal of Crystal Growth, Elsevier), p. 51-56.

[5] Beecher S J, Grant-Jacob J A, Parsonage T L, 2016 *Yb: YAG planar waveguide lasers grown by pulsed laser deposition: 70% slope efficiencies at 16 W of output power* (Solid State Lasers XXV: Technology and Devices. International Society for Optics and Photonics 9726) 97261Z

[6] Brown M S, Arnold C B 2010 *Fundamentals of laser-material interaction and application to multiscale surface modification* (Laser precision microfabrication. Springer, Berlin, Heidelberg) p 91-120.

[7] Bonefačić I, Blecich P. 2015 Two-color temperature measurement method using BPW34 PIN photodiodes (Engineering Review: *Međunarodni časopis namijenjen publiciranju originalnih istraživanja s aspekta analize konstrukcija, materijala i novih tehnologija u području strojarstva, brodogradnje, temeljnih tehničkih znanosti, elektrotehnike, računarstva i građevinarstva*) chapter 35(3) p 259-266.

[8] Tapetado A, Díaz-Álvarez J, Miguélez M H 2016 Two-color pyrometer for process temperature measurement during machining (Journal of lightwave technology) chapter 34(4) p1380-1386

[9] Andrew W 1994 Handbook of deposition technologies for films and coatings (Elsevier)

[10] Brailovsky A B, Gaponov S V, Luchin V I. 1995 Mechanisms of melt droplets and solid-particle ejection from a target surface by pulsed laser action (Journal of Applied Physics) 61(1) p 81-86

Journal of Physics: Conference Series **2274** (2022) 012006 doi:10.1088/1742-6596/2274/1/012006

[11] Prentice J, Grant-Jacob J, Kurilchik S, Mackenzie J I, Eason R.W, 2019 Particulate reduction in PLD-grown crystalline films via bi-directional target irradiation (Journal of Applied Physics, SPRINGER)

[12] Kurilchik. S. V., Grant-Jacob J. A., Prentice J., Hua P., Mackenzie J.I., and Eason R.W., "Pulsed-laser-deposited Yb:YAG planar-waveguide amplifier", in 8th EPS-QEOD Europhoton Conference, Spain, (2018).

[13] Li Z, Liu B, Wang J 2012 Mechanism of Intrinsic Point Defects and Oxygen Diffusion in Yttrium Aluminum Garnet: First-Principles Investigation (Journal of the American Ceramic Society) 95(11) p 3628-3633

[14] Ismail R A, Rasheed B G, Salm E T 2007 *Transparent and conducting ZnO films prepared by reactive pulsed laser deposition* (Journal of Materials Science: Materials in Electronic) 18(4) p 397-400.

[15] Boles S. 2016 Development of Traceable Capabilities in Non-Contact Thermal Metrology (Technological University Dublin)

[16] Howell J R, Menguc M P, Siegel R 2015 *Thermal radiation heat transfer* (CRC Press)

[17] Fang Tiegang 2002 An extension of Wien's displacement law for blackbody radiation (Elsevier Science) 29(6):757-761

[18] J. Marr, F. Wilkin 2011 A better presentation of Planck's radiation law (American Journal of Physics) 80(5) 10.1119/1.3696974

[19] Pason O, Grebe-Ellis J 2017 Planck's radiation law, the light quantum, and the prehistory of indistinguishability in the teaching of quantum mechanics (European Journal of Physics) 38(3):035404

[20] Van Vleck J H. 1924 The absorption of radiation by multiply periodic orbits, and its relation to the correspondence principle and the Rayleigh-Jeans law. Part I. Some extensions of the correspondence principle (Physical Review), 24(4) p 330

[21] Barrington S J, Eason R W 2000 Homogeneous substrate heating using a CO2 laser with feedback, rastering, and temperature monitoring (Review of Scientific Instruments) 71(11) p 4223-4225

[22] De Vos J C. 1954 *Evaluation of the quality of a blackbody* (Physica) 20(7-12) p 669-689

[23] May-Smith T C, Muir A C, Darby M S B,2008 Design and performance of a ZnSe tetra-prism for homogeneous substrate heating using a CO2 laser for pulsed laser deposition experiments (Journal of Applied optics) 47(11) p 1767-1780.

Journal of Physics: Conference Series **2274** (2022) 012007 doi:10.1088/1742-6596/2274/1/012007

Spectral characteristic of symmetric double tapered plastic optical fibers

M. Marinah[1, *], I. Puspita[1], D.P.M. Banurea[1], Zyrlirosa[1], A.M. Hatta[1], and Sekartedjo Koentjoro[1]

[1]Department of Engineering Physics, ITS, Indonesia

[*]Corresponding author: murtimarinah7@gmail.com

Abstract. Plastic optical fiber (POF) sensor has been immensely developed for many sensing applications including physical and chemical/bio sensor. Since it offers the advantages of non-destructive, easy fabrication and installation, immunity to electromagnetic interference, compact, low fabrication costs and higher durability compare to silica fiber sensor. Tapered POF is one of the modified structures on POF by reducing its diameter and induces guided modes in the core to be evanescent modes. As a sensor, the tapered structure on POF increase the sensitivity of non-tapered POF. The number of tapers on optical fiber affect its characteristics as well. Hence, it is important to investigate the characteristic of tapered POF with the variation on number of tapers. In this paper, the spectral characteristic of symmetric double tapered POF (DT POF) was investigated. The fabrication of symmetric DT POF was conducted by reducing the core diameter using chemical process as described in. The core diameter and the length of the tapered were fixed at 0.45 mm and 1 cm, respectively. The distance between two tapers was varied from 1-3 cm. Transmission spectrum characterization of symmetric DT POF was measured at near infrared wavelength range (900-1600 nm) using a spectrometer by OTO-Photonics.

1. Introduction

Refractive index measurement has an important role when dealing with quality check of certain materials used in different industries, such as biomedicine, chemical, food safety, pharmaceutical, etc. There are many methods to measure refractive index based on optical fiber probes, such as Optical Fiber Grating [1–3], Surface Plasmon Resonance (SPR) [4,5], Singlemode-Multimode-Singlemode fiber [6–8], Mach Zehnder Interferometer [9,10], and optical fiber Fabry-Perot cavity [11,12]. Plastic optical fiber has great potential to be used as a refractive index sensor. Since it offers the advantages of non-destructive, easy fabrication and installation, immunity to electromagnetic interference, compact, low fabrication costs and higher durability compared to silica fiber sensor [13].

Plastic optical fiber (POF) sensors have been widely developed and structurally modified to improve sensor performance, such as increasing sensitivity. Taper structure is the most effective structure for reducing the plastic optical fiber diameter of all existing fiber optic structures [14]. Therefore, tapered POF can induce the guided mode in the core into evanescent modes [15] thereby increasing the interaction between the evanescent modes in the plastic optical fiber and the surrounding environment [14]. The increasing of evanescent waves interaction with the environment can increase the sensor's sensitivity. Interaction between the evanescent wave with the environment results an optical transmission attenuation, which can be used to measure the refractive index of the external environment sensitively. Some previous studies have proven that tapered POF have great potential for bio/chemical sensors and their sensitivity can increase according to the geometry of the structure [16–19], including the number of tapers. The number of tapers on plastic optical fiber affect its characteristics as well. Hence, it is important to investigate the characteristic of tapered POF with the variation on number of tapers. This paper demonstrates a symmetric double tapered (DT) structure as a refractive index sensor based on plastic optical fiber. The refractive index of the external environment is measured by detecting the wavelength shift of the output transmission spectrum.

2. Methods

2.1. DT POF Principle

The proposed sensing system consist of two tapers region that are separated with various distances of 1, 2, and 3 cm. When light is transmitted into the first taper area, the core mode is excited to high-order mode in the cladding area and propagated along the cladding [20]. The guided high-order mode that penetrates into the cladding layer produces a strong evanescent field between the cladding surface and the external environment [21]. The guided and evanescent modes will propagate along the taper length (L_{taper}). The evanescent modes that have different incident angles cause interference within modes in cladding first tapered area. The cladding and core modes will be coupled back into un-tapered region from the up-taper of first taper area and result an interference spectrum (shown in Figure 1). Due to the different geometric structures in the un-tapered region between two taper structures, coupling between modes from first tapered is highly possible occurred in the core [22]. Those modes propagate with different optical path lengths and as a result, they interfere with each other along the L propagation direction [22]. At the end point of the un-tapered region, the intensity distribution was formed not only in the middle of core diameter but also in the whole area of core diameter [23]. Then the modes will be coupled back in the second taper structure and experience the same phenomenon as the first taper. Therefore, in the DT POF structure, evanescent modes are formed not only in the first taper region but also in the second taper region. This yields the more interaction between evanescent modes and the sample.

2.2. DT POF Fabrication

The tapered structure was seen after reducing the core diameter through chemical etching as described in [17]. The DT POF was fabricated with two tapers structures separated with different distance lengths (L) from 1 to 3 cm, and the taper waist diameter and taper length of 0.45 mm and 1 cm respectively as in Figure 1.

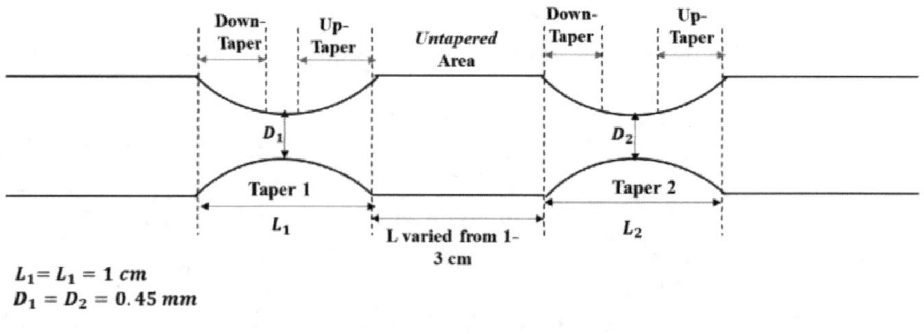

Figure 1. A schematic structure of DT Structure

The plastic optical fiber (POF) used in this research is the Super EskaTM, Step Index Multimode POF with core and cladding diameter of 980 μm and 20 μm, respectively. The core of DT POF has a refractive index of 1.49 and a numerical aperture of 0.50. Figure 2 depicts a microscope image of the DT POF. One can see that the taper structure 1 and taper structure 2 are identical and have the same taper waist diameter of 0.45 mm.

Journal of Physics: Conference Series **2274** (2022) 012007 doi:10.1088/1742-6596/2274/1/012007

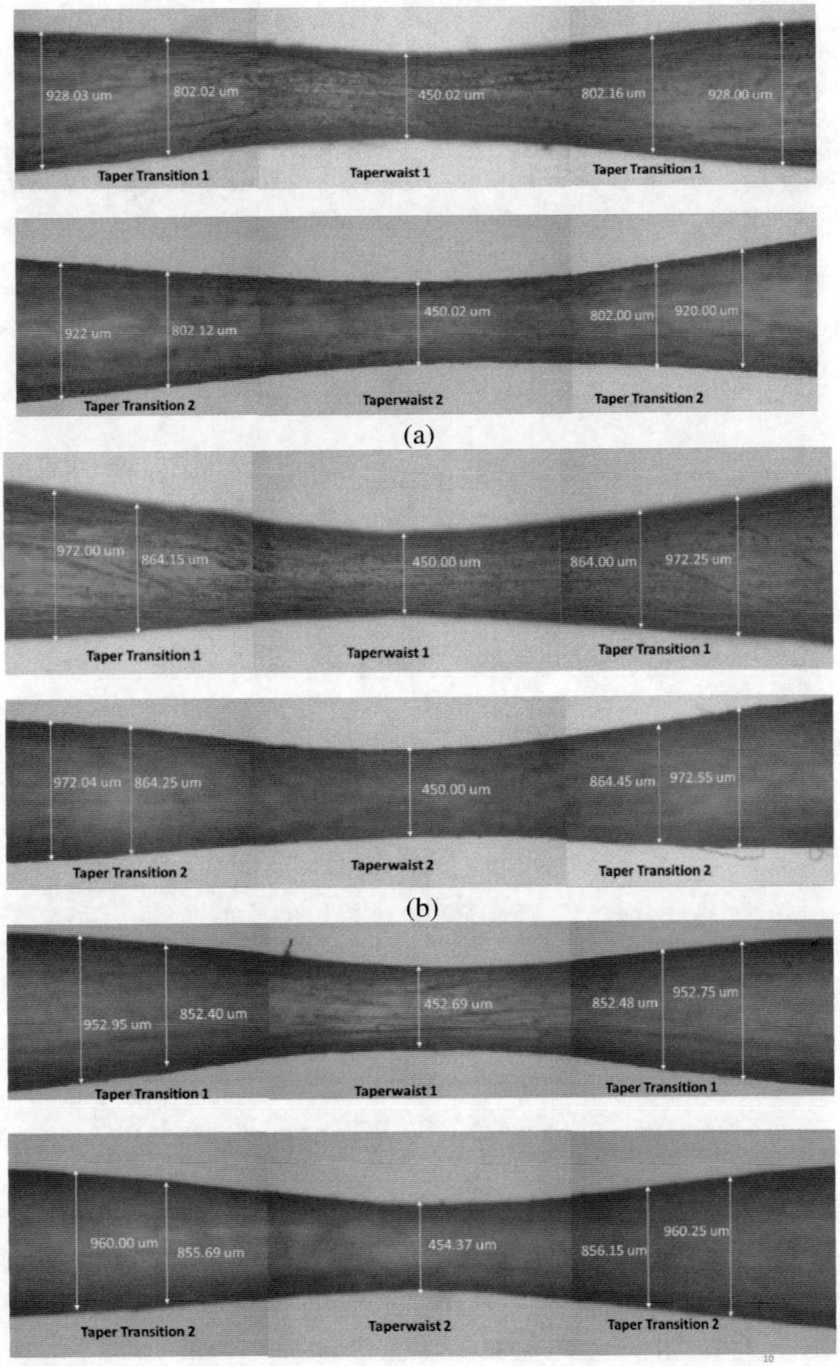

(c)

Figure 2. Microscopic image of symmetric DT POF with distance between two tapers (a) 1 cm (b) 2 cm (c) 3 cm

2.3. Experimental Set-Up

The experimental set-up is presented in Figure 3. The halogen lamp (Ocean Optics HL-2000-HP) was used as the light source with the operating wavelength 350-1700 nm. The DT POF connected to a spectrometer (NIR Oto SW-2560) with the operating wavelength of 900-1600 nm. The spectrometer is connected to the computer to record the output measured spectrum. Data was recorded while the DT probe immersed in the sample. Measurements were repeated 5 times according to the Federer formula [24], in order to determine the precision of the measurement conducted by DT POF.

In our experiment, we use various vegetables oil as in Table 1. In addition, we also use lard with concentration of 0-5% (v/v) with an increment of 1% mixed with olive oil as samples as described in a previous study [24]. Lard was obtained by extracting pork belly fats which are purchased from the local market. Pork belly fats were heated up until they melted to extract the lard. The extracted lard was filtered by using a paper filter. Then, lard and olive oil are mixed by using an ultrasonic bath (Krisbow, KLS-303363) in 30 minutes to get the homogeneous mixture.

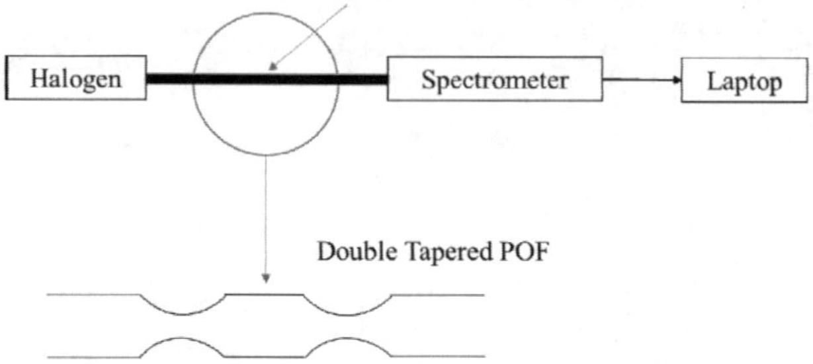

Figure 3. Set-up of experiment

3. Experiment Result and Discussion

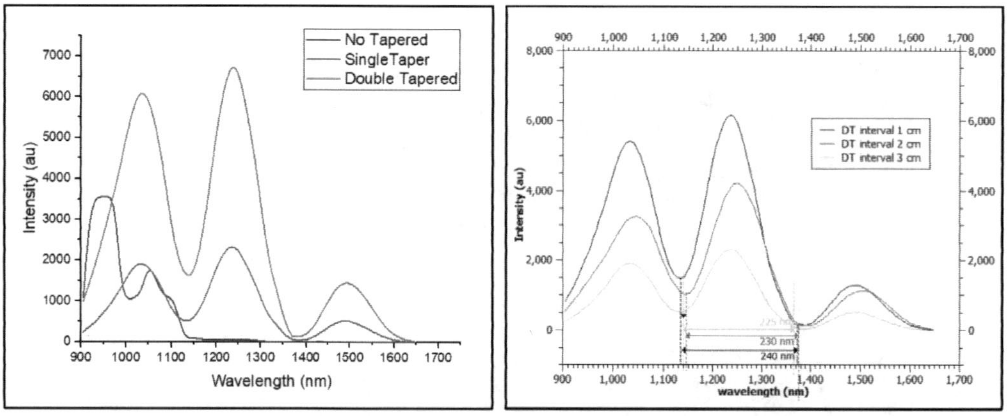

(a) (b)

Figure 4. Spectrum Transmission of Symmetric DT POF without Sample (a) Compared with Other Structures (b) with Their Respective Free Spectral Range (FSR) Values

The transmission spectra of the tapered POF structures are presented in Figure 4. One can see that the tapered structure, both single and double tapered induced the modal interference phenomenon as indicated by the appearance of peaks. The interference spectrum only appear when the tapered structure in non-adiabatic condition [20]. In non-adiabatic conditions, the core mode will be excited to the higher order mode due to the geometry of the tapered structure in down-taper transition[20]. These high-order modes will be excited to the cladding area with different angles of incidence and become evanescent modes. Evanescent modes with different angles of incidence allow them to interfere with each other in the taper area. Evanescent modes along with core modes will be coupled again in the up-tapered transition, hence result the interference in output spectrum of Single Taper. In DT POF structure, this phenomenon occurs in both tapered areas. The modes from first tapered will be propagate in un-tapered region with varied incidence angle, hence allow them to interfere with each other in un-tapered region. Due to the different geometric structures in the un-tapered region between two taper structures, coupling between those modes is highly possible occurred in the core [22]. Then the modes will be coupled back in the second taper structure and experience the same phenomenon as the first taper. Therefore, the output transmission of DT POF also shows the interference spectrum.

The transmission spectrum of a non-adiabatic tapered arises the interference of modes excited in the taper, as well as the absorption spectrum of the surrounding medium [25,26]. At the output spectrum, the distance between two peaks or dips can be attributed by a Free Spectral Range (FSR). The FSR decreases as the distance between two tapers increases as shown in Figure 4(b). The results show agreement with the theory as in [4]. The change of FSR value indicates a change in the interference spectrum [5]. It caused by the change of propagation modes in fiber. According to the Lambert Beer Law (Equation 1), the chemical composition of the surrounding medium influences the propagation of the evanescent modes and thus the propagating modes [25,26]. Hence, the change of refractive index will also influence the propagation of modes and result the difference of interference output spectrum. Therefore, by observing the change in FSR of the output spectrum, RI can be determined.

$$\log\frac{I}{I_0} = c\alpha L \tag{1}$$

where I and I_0 are the light intensities; c is the concentration of chemical compound, L is the optical path length and α is the molar extinction coefficient. Based on the theory in [4], the change of FSR is not only influenced by the changes in refractive index of surrounding medium but also influenced by the tapered geometry (in this experiment, the change of tapered geometry is the varied of distance between the two tapers).

We started with the various vegetable oil samples before investigating the adulterated olive oil samples. Table 1 shows the refractive index of various vegetable oils used in this experiment. The output spectrum of each DT POF structures (depict in Figure 6) was evaluated by observing the change of the FSR peak (ΔFSR_{peak}) and the FSR dip (ΔFSR_{dip}). The result is depicted in Figure 5.

Table 1. Refractive Index Vegetable Oil

Vegetable Oil	Refractive Index
Coconut Oil	1.455
Palm Oil	1.465
Olive Oil	1.467
Canola	1.47
Sunflower	1.473

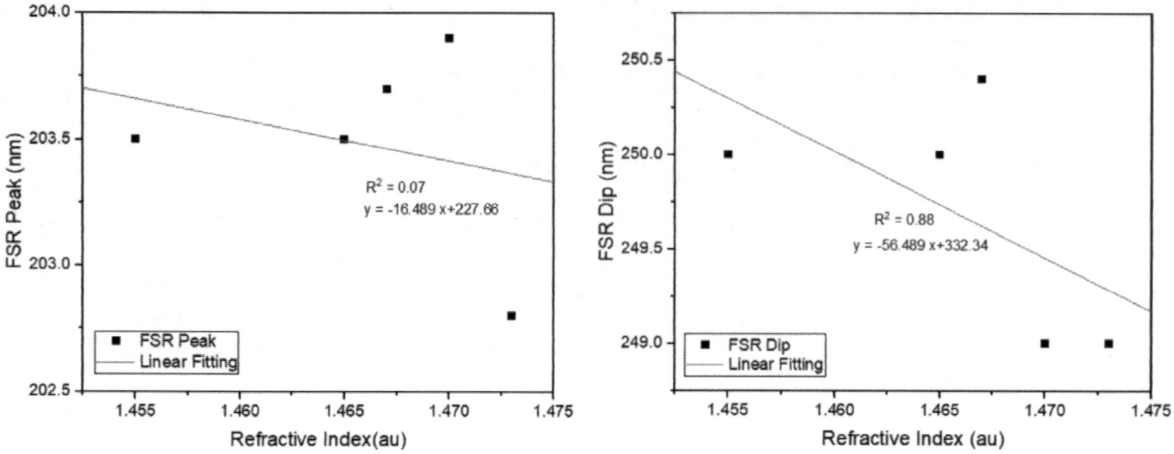

(a)

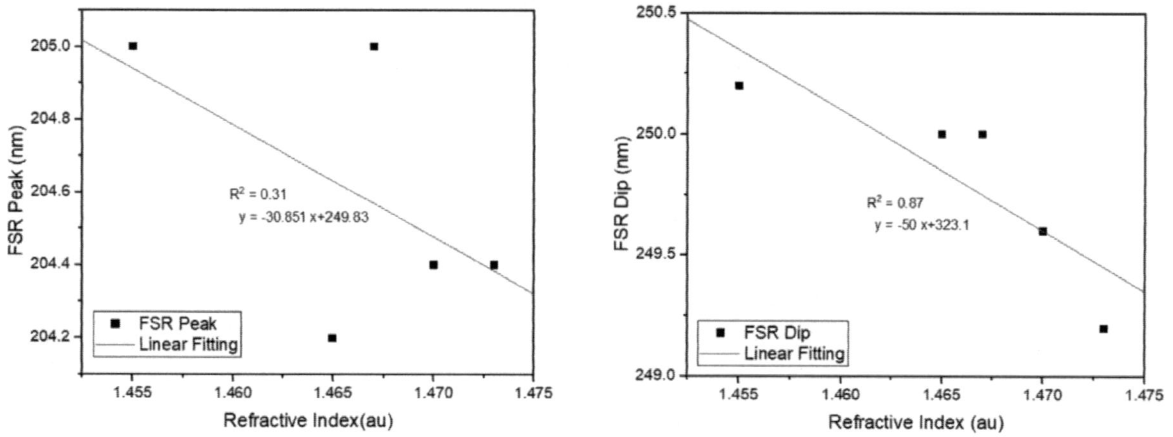

(b)

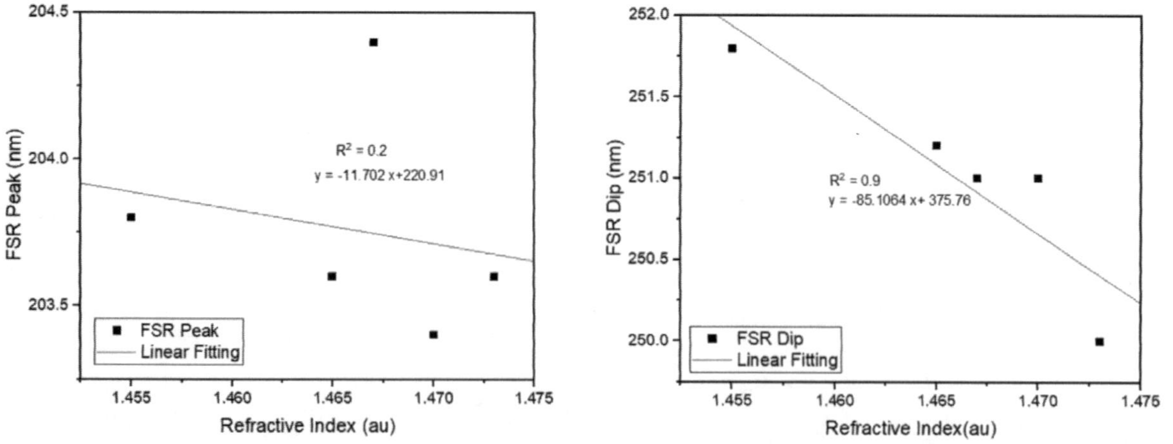

(c)

Figure 5. The changes in FSR value as refractive index increases of DT POF structures with the distance between two tapers (a) 1 cm (b) 2 cm (c) 3 cm

One can see in Figure 5(a) - 5(b), the ΔFSR_{peak} in DT POF structures with the distance between two tapers of 1 cm, 2 cm, and 3 cm, have poor linearity indicated by the low correlation coefficient (R^2)

of 0.07, 0.31, and 0.2 respectively. Meanwhile the ΔFSR_{dip} in DT POF structures with the distance between two tapers of 1 cm, 2 cm, and 3 cm, have a good linearity with higher correlation coefficient than ΔFSR_{dip}, which are 0.88, 0.87, and 0.9 respectively. In addition, ΔFSR_{dip} has higher sensitivity than ΔFSR_{peak} indicated by the value of gradient in linear regression, hence the dip wavelength area is an appropriate area for refractive index measurements. Therefore, the wavelength shifting was evaluated and observed in the dip wavelength of the output transmission spectrum (depicted in Figure 6).

The wavelength shifting of dip B due the increase of refractive index is more shifted sequentially than dip A as depicted in Figure 6 (a) – Figure 6 (c). It indicates that the change of refractive index is more sensitively observed in dip B. The linearity and sensitivity of dip B are depicted in Figure 7. One can see in Figure 7, the dip wavelength shifted toward shorter wavelength (*blue shifting*) in all DT POF structures. In addition, from the Figure 7, the DT POF with the distance between two tapers of 3 cm has a good linearity of 0.93 and the highest sensitivity of 355.319 nm/au compared with the others. Then, we try with other samples by adding lard (0-5%) into olive oil to make sure the results that have been obtained.

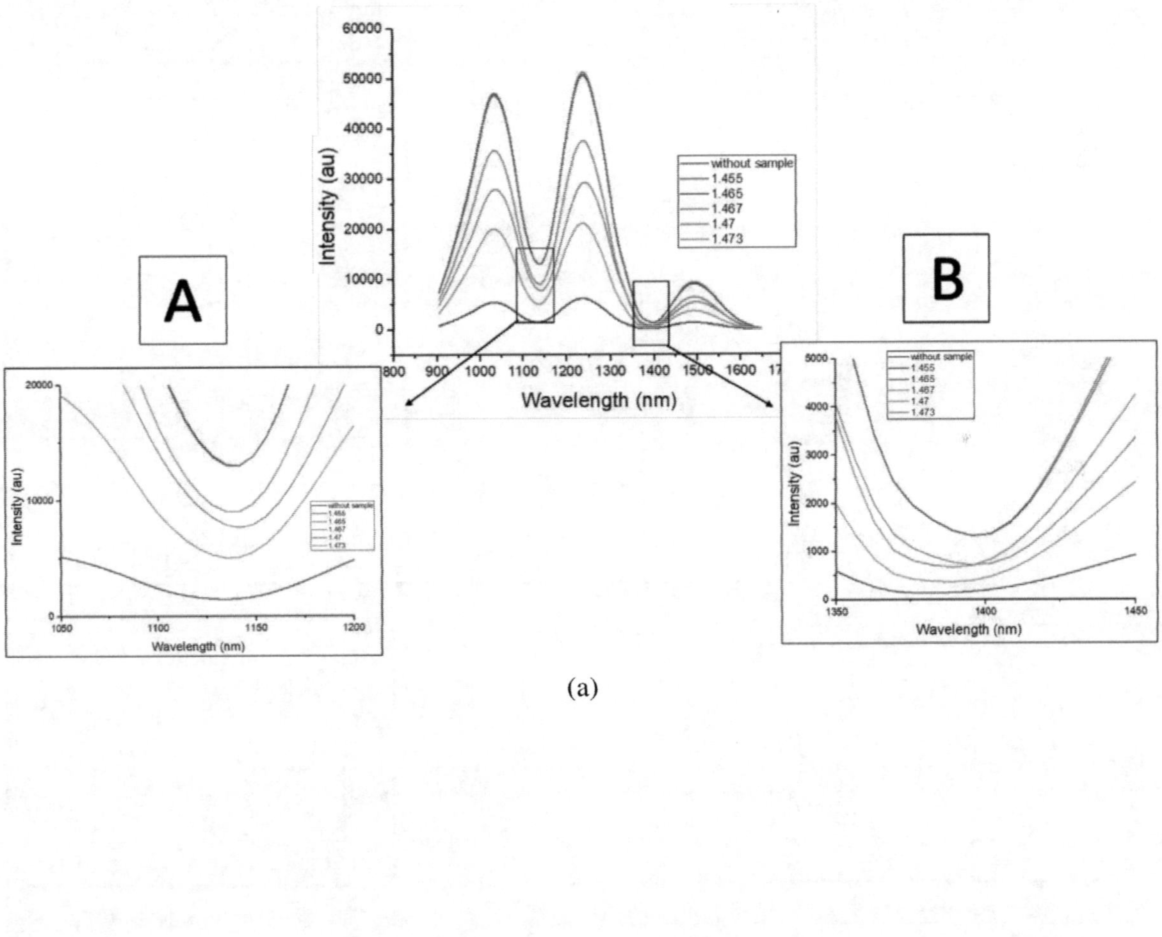

(a)

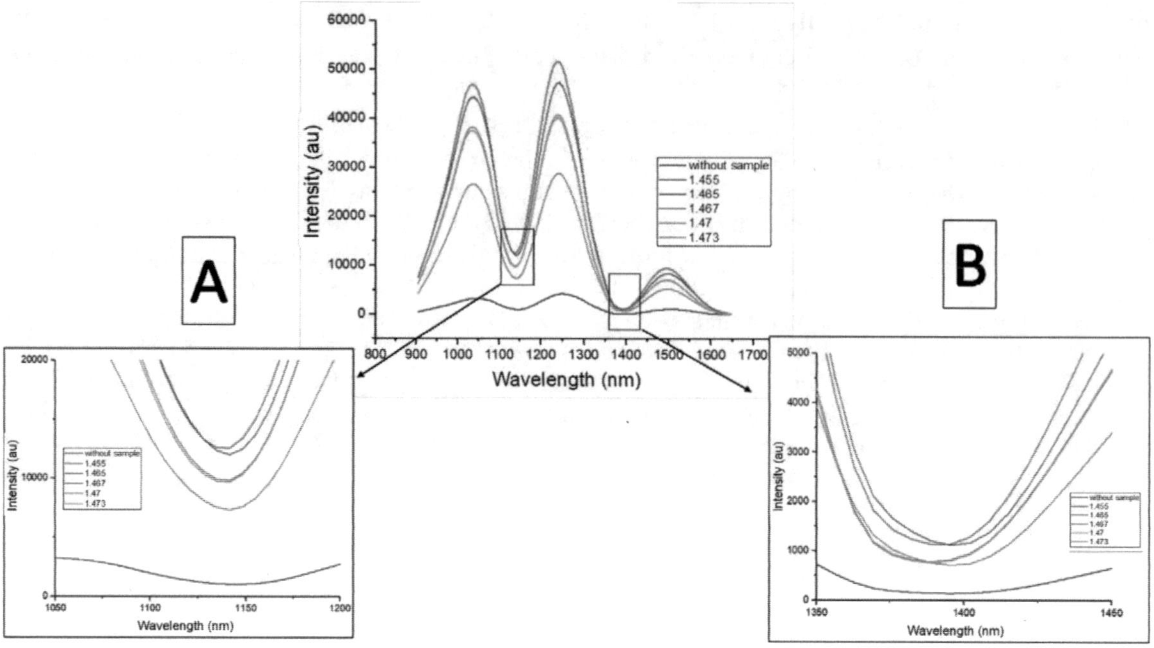

(b)

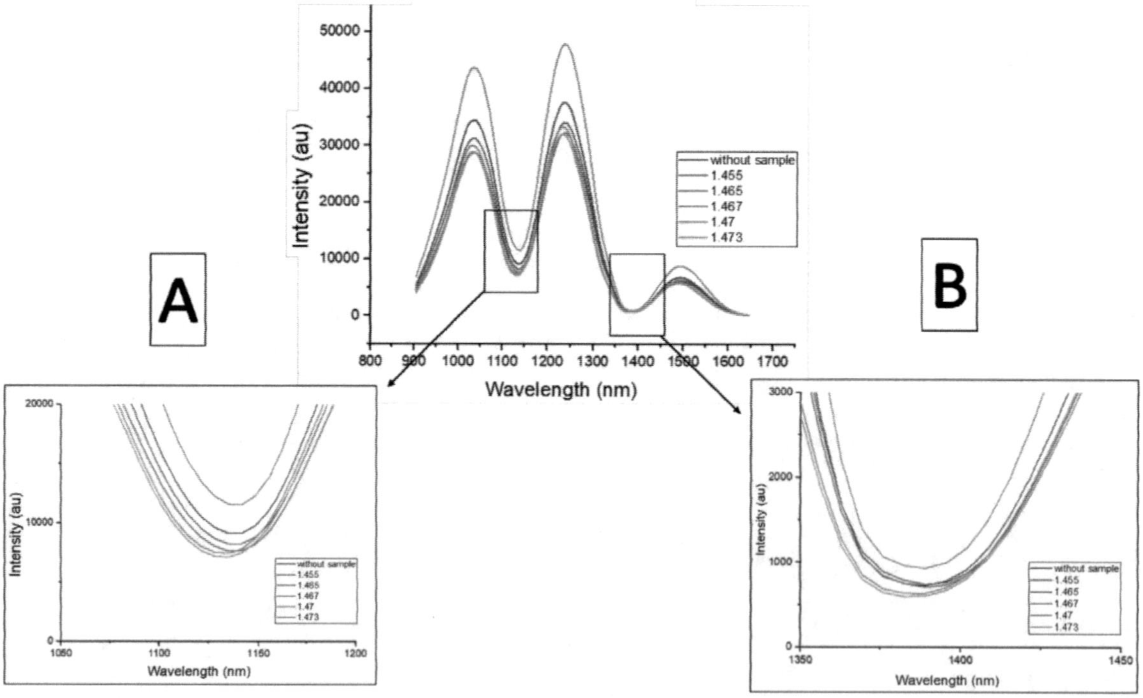

(c)

Figure 6. Dip wavelength shifting of DT POF (a) L=1 cm (b) L=2 cm (c) L=3 cm

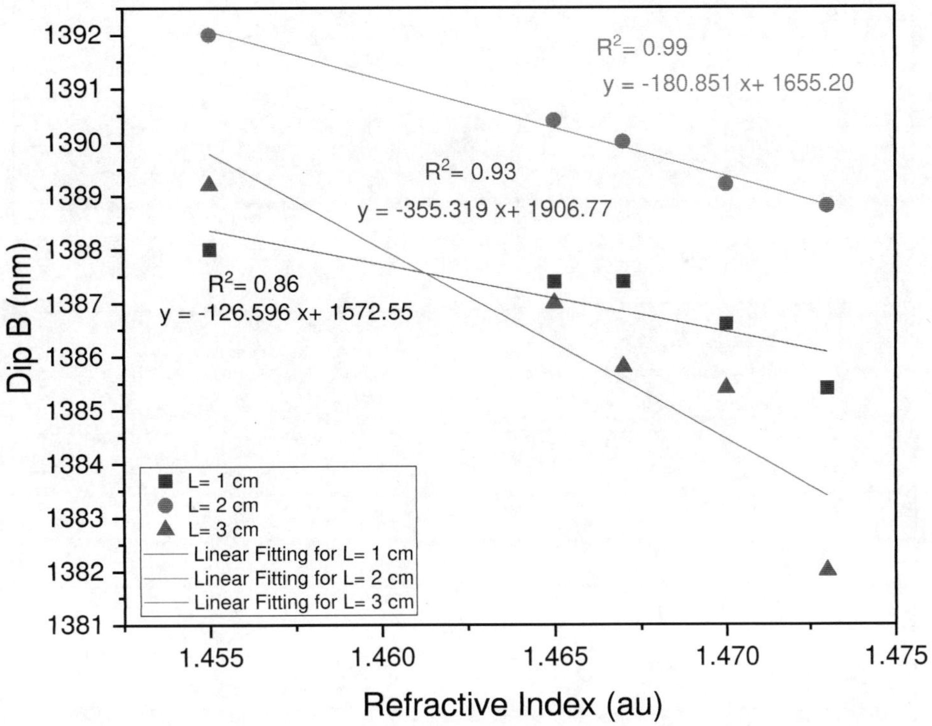

Figure 7. Plot of Dip Wavelength Shift due to increase of refractive index

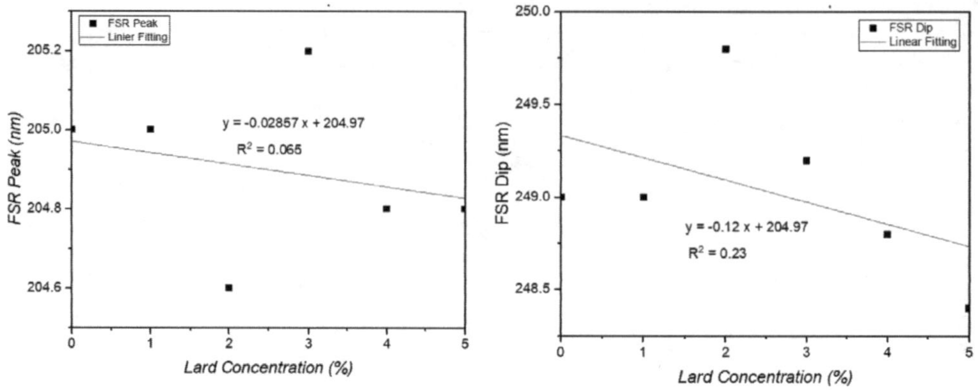

(a)

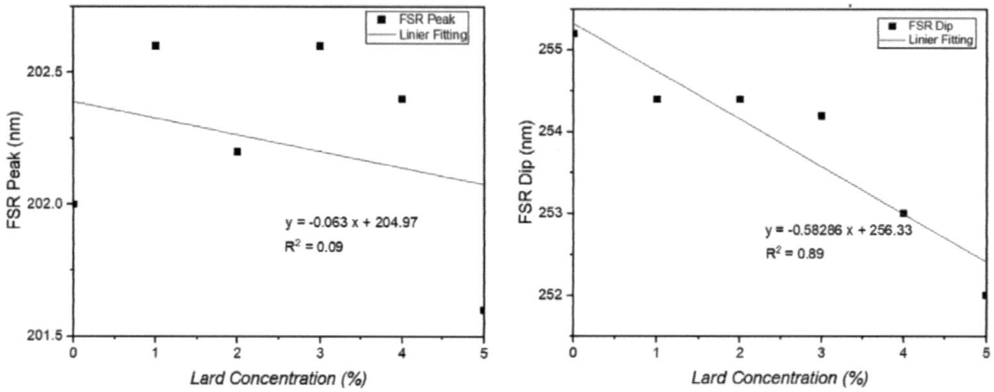

(b)

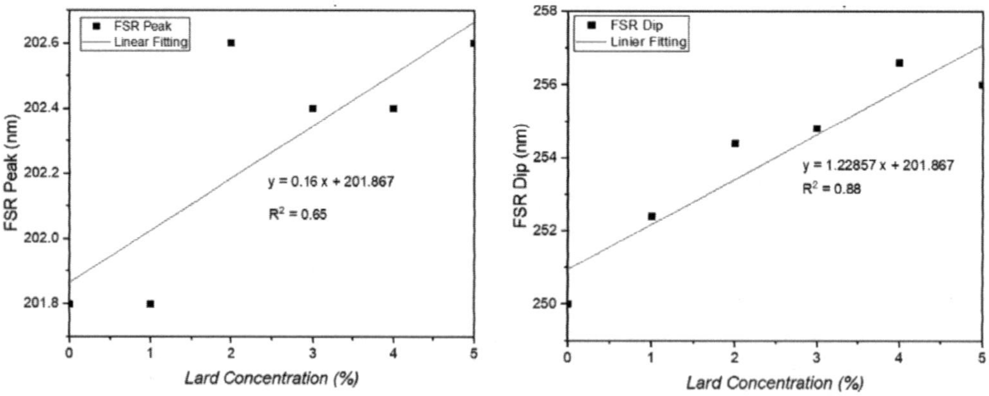

(c)

Figure 8. The changes in FSR value as concentration of lard in olive oil increases (a) DT POF interval 1 cm (b) DT POF interval 2 cm (c) DT POF interval 3 cm

In the next experiment, the olive oil adulterated with 0-5% (v/v) of lard, were used as the sample. The refractive index of the sample is increased as the increase of lard concentration in olive oil, as in previous work [24]. The output spectrum transmission of the measurement by each DT POF structures depicted in Figure 9. The change of FSR value, both in dip and peak of the output spectrum transmission were evaluated in Figure 8. One can see in Figure 8(a) – 8(c), the ΔFSR_{peak} in DT POF structures with the distance between two tapers of 1 cm, 2 cm, and 3 cm, have very poor linearity indicated by the low correlation coefficient (R^2) of 0.065, 0.09, and 0.65 respectively. The ΔFSR_{dip} in DT POF structure with the distance between two tapers of 1 cm also has poor linearity of 0.23 but still higher than the ΔFSR_{peak} linearity. Meanwhile the ΔFSR_{dip} in DT POF structures with the distance between two tapers of 2 cm, and 3 cm have good linearity, which are 0.89, and 0.88 respectively. In addition, ΔFSR_{dip} in DT POF structures with the distance between two tapers of 2 cm and 3 cm have higher sensitivity than ΔFSR_{peak} indicated by the value of gradient in each linear regression. It indicates that the dip wavelength area is an appropriate area for refractive index measurements. Therefore, the dip wavelength shift was observed in the output transmission spectrum shown in Figure 9. The linearity and sensitivity of dip A dan dip B from each structure are depicted in Figure 10.

In Figure 9 and Figure 10, the dip wavelength shifted toward shorter wavelength (*blue shifting*) in the DT POF structure with the distance between two tapers of 1 cm and 2 cm. Meanwhile, the dip wavelength for DT POF structure with the distance of 3 cm shifted toward longer wavelength (*red shifting*). This may be caused by the inclination angle of the fabricated up-taper and down-taper transition were not really symmetric or there is a delta as far as $\pm\theta$. In a previous work [27], the inclination angle difference between the up and down transition of tapered POF sensor was specifically investigated and it obtained the reflected optical response. It was found in Figure 10 that the sensitivity gets higher with the increase of the distance between two tapers and this result show an agreement with the result of vegetable oil samples where the highest sensitivity is obtained by the DT POF structure with the distance between two tapers of 3 cm. Other parameters of sensor performance were evaluated in Table 2.

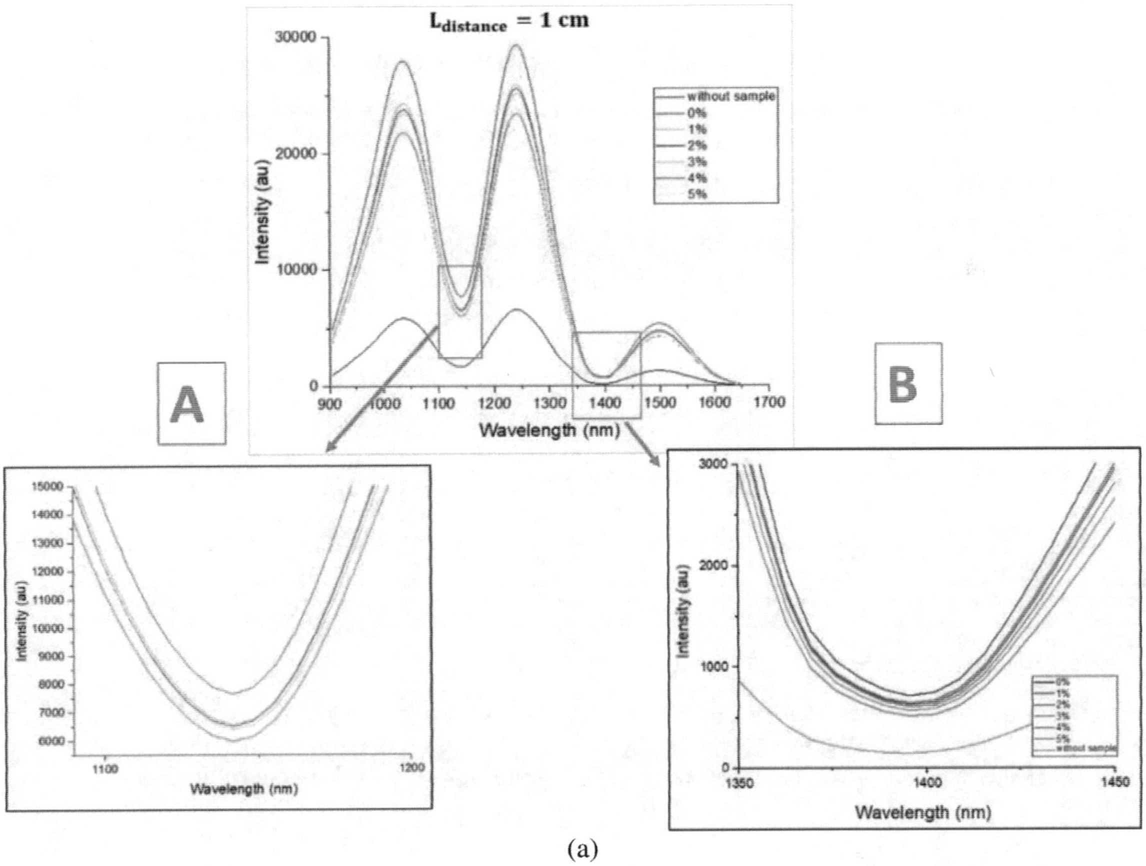

(a)

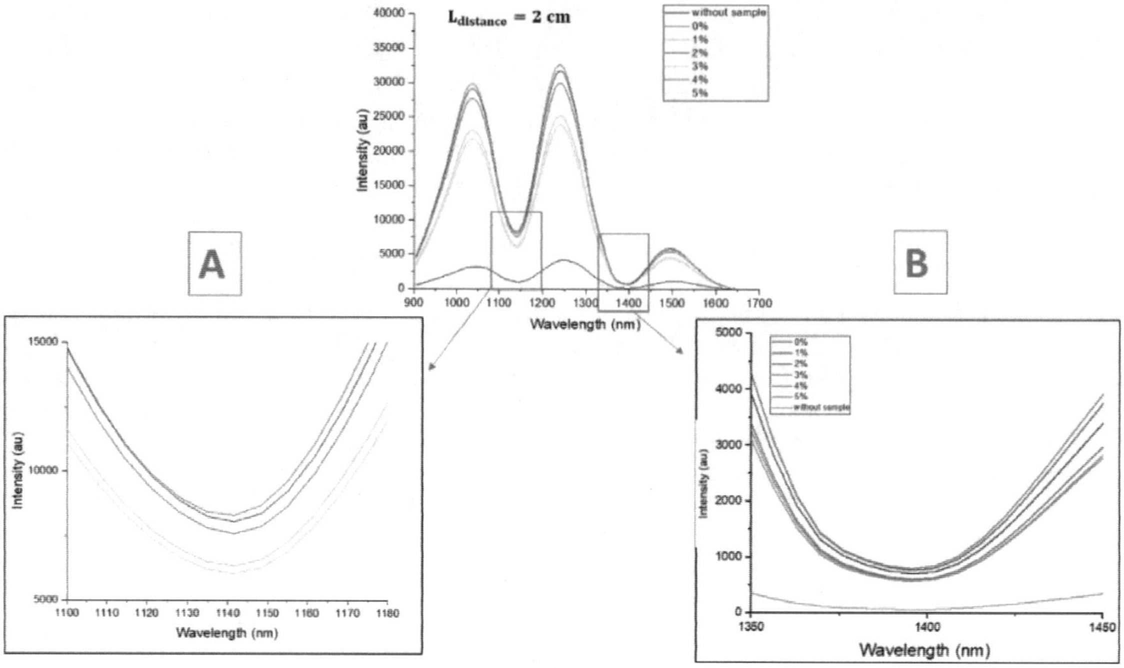

(b)

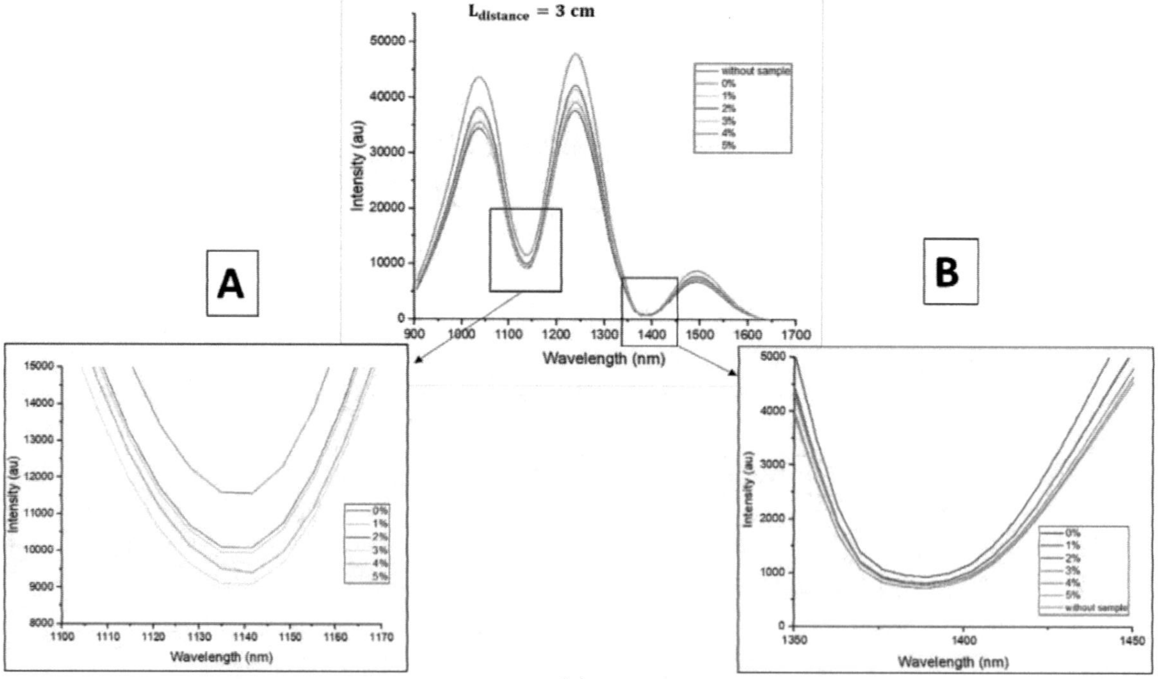

(c)

Figure 9. Dip wavelength shifting of DT POF (a) L=1 cm (b) L=2 cm (c) L=3 cm

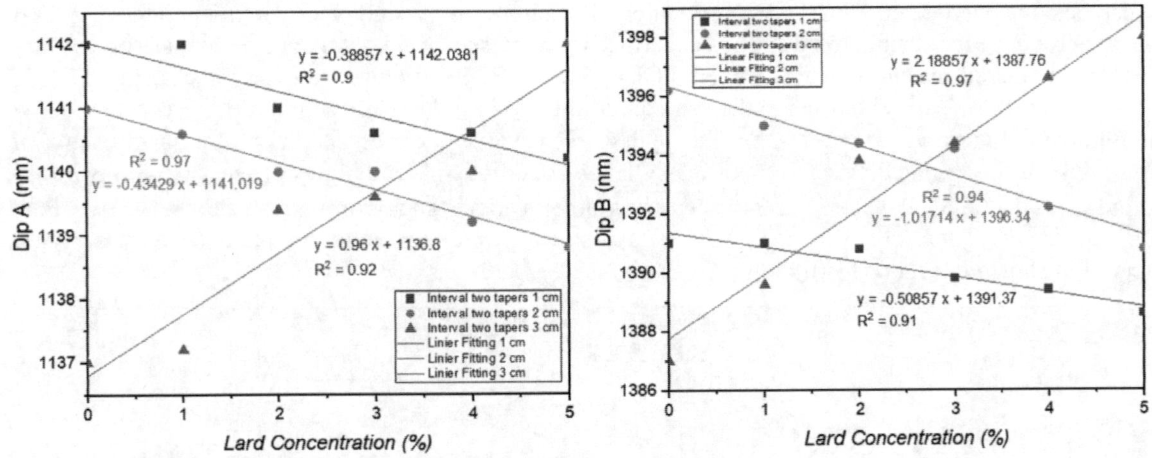

Figure 10. Plot of Dip Wavelength Shift due to increase in lard concentration

Sensor as the first element in measurement system gives an output signal as a function of the input of a certain physical quantity, thus beside of sensitivity, the Limit of Detection is the important parameter in sensor performance. Limit of detection is the smallest limit of an analyte that can still be detected by the sensor [28], thus the smaller limit of detection value, the better the sensor performance. Based on Table 2, limit of detection gets smaller as the distance between the tapers increases. Limit of detection calculated by the following equation [29,30] with D is the standard deviation and S is the sensitivity:

$$\text{Limit of detection} = \frac{3D}{S} \tag{2}$$

Table 2. Performance evaluation of DT POF

Parameters	Dip A			Dip B		
	DT POF interval 1 cm	DT POF interval 2 cm	DT POF interval 3 cm	DT POF interval 1 cm	DT POF interval 2 cm	DT POF interval 3 cm
Linear Range (%) (v/v)	0-5	0-5	0-5	0-5	0-5	0-5
Span Output (nm)	1.8	2.2	5	2.4	5.4	11
Sensitivity (nm/%)	0.3886	0.4343	0.96	0.5086	1.0171	2.1886
Linearity (%)	90	97	92	91	94	97
Standard Deviation (nm)	0.22	0.20	0.11	0.23	0.27	0.18
Limit Of Detection (%)	1.70	1.38	0.34	1.36	0.80	0.25

Table 3. Precision of the DT POF

Distance between two tapers	Relative Standard Deviation	
	Dip A	Dip B
1 cm	0.0006	0.0003
2 cm	0.0006	0.0006
3 cm	0.0002	0.0002

A sensor has a good precision if it has the relative standard deviation value of less than 2% [31]. Table 3 shows the relative standard deviation value of the DT POF in all variations of less than 2%, so it can be concluded that the DT POF has good precision in refractive index measurement hence DT POF sensors have great potential to be applied as refractive index sensor. In obtained the comparison between the single tapered plastic optical fiber (ST POF) with DT POF structure, an experiment using ST POF structure was also carried out using the samples of olive oil adulterated with lard 0-5% (v/v) as shown in Figure 11. Figure 11 (a) depicts the change in the FSR value due to the addition of lard concentration in sample. It shows that the ΔFSR_{peak} of ST POF has good linearity and higher sensitivity than ΔFSR_{dip}. It indicates that in ST POF structure, the peak wavelength area is an appropriate area for refractive index measurements. Therefore, the peak wavelength shifting was observed in the output transmission spectrum shown in Figure 11 (b).

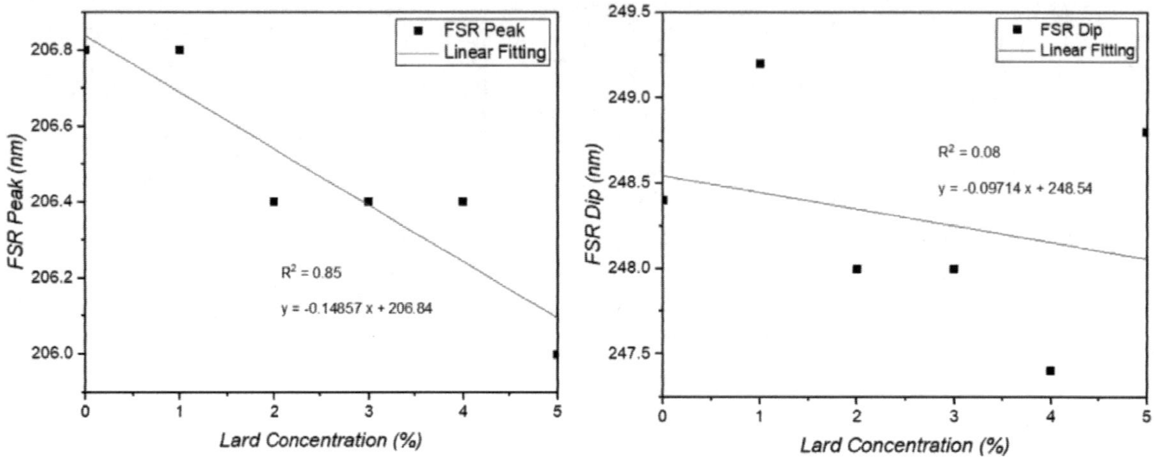

(a)

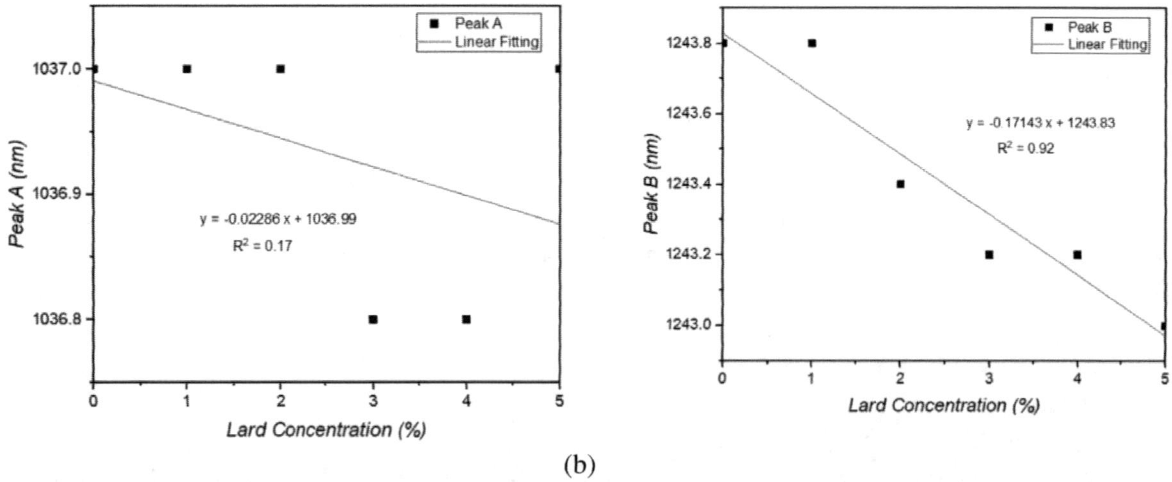

(b)

Figure 11. (a) The changes in FSR value as concentration of lard in olive oil increases (b) Peak Wavelength Shift due to increase in lard concentration

One can see in Figure 11 (b) the peak wavelength shifting in area B is more sensitive and linear than area A. Therefore, we compared the measurement of the wavelength shifting in Peak B of ST POF with the wavelength shifting in Dip B of DT POF structure as shown in Table 4. Table 4 shows that the DT POF has higher sensitivity than the ST POF structure. This is due to the fact that more evanescent modes occur in the DT POF than the ST POF structure due to the addition of a second taper. Therefore, the interaction of the evanescent mode with the samples in DT POF structure will be higher than the ST POF structure. This, in effect, result the higher transmission power loss in DT POF structure. This result shows an agreement with previous study regarding the comparison between DT POF and ST POF structure [15].

Table 4. Performance comparation between the DT POF and ST POF structure

Parameters	DT POF			ST POF
	interval 1 cm	interval 2 cm	interval 3 cm	
		Dip B		Peak B
Linear Range (%) (v/v)	0-5	0-5	0-5	0-5
Span Output (nm)	2.4	5.4	11	0.8
Sensitivity (nm/%)	0.5086	1.0171	2.1886	0.1714
Linearity (%)	91	94	97	92
Standard Deviation (nm)	0.23	0.27	0.18	0.17
Limit Of Detection (%)	1.36	0.80	0.25	2,98

4. Conclusions

Based on the result of the experiment, it can be concluded that the DT POF sensor has great potential to be used as a refractive index (RI)-based sensor. The RI sensing principle of DT POF structure is utilizing spectral information in the form of changes in the interference spectrum that occur due to the change of effective refractive index (ERI) and geometric characteristics of the DT POF sensor. The sensitivity of DT POF sensor is higher than ST POF because there are more interactions of the evanescent mode with the sample in the DT POF structure than the ST POF due to the addition of a second taper. The DT POF sensor has good linearity and good precision to measure the refractive index of sample. Moreover, the limit of detection (LOD) characteristic will get smaller as the distance between two tapers increase. In opposite with LOD, the sensitivity of the DT POF sensor will get higher as the distance between two tapers increase. Thus, indicate that the performance of DT POF structure will be greatly influenced by its geometric parameters such as the length between tapers.

References

[1] Tsigaridas G, Polyzos D, Ioannou A, Fakis M and Persephonis P 2014 Theoretical and experimental study of refractive index sensors based on etched fiber Bragg gratings *Sensors Actuators, A Phys.* **209** 9–15

[2] Geng T, Zhang S, Peng F, Yang W, Sun C, Chen X, Zhou Y, Hu Q and Yuan L 2017 A Temperature-Insensitive Refractive Index Sensor Based on No-Core Fiber Embedded Long Period Grating *J. Light. Technol.* **35** 5391–6

[3] Xue P, Yu F, Wu B, Bao H and Zheng J 2020 Investigation of a D-Shaped Plastic Optical Fiber Assisted by a Long Period Grating for Refractive Index Sensing *IEEE Sens. J.* **20** 842–7

[4] Zhang Y, Liu M, Zhang Y, Liu Z, Yang X, Zhang J, Yang J and Yuan L 2020 Simultaneous measurement of temperature and refractive index based on a hybrid surface plasmon resonance multimode interference fiber sensor *Appl. Opt.* **59** 1225

[5] Li H, Qian X, Zheng W, Lu Y, Siyu E and Zhang Y nan 2020 Theoretical and experimental characterization of a salinity and temperature sensor employing optical fiber surface plasmon resonance (SPR) *Instrum. Sci. Technol.* **48** 601–15

[6] Xue L L and Yang L 2012 Sensitivity enhancement of RI sensor based on SMS fiber structure with high refractive index overlay *J. Light. Technol.* **30** 1463–9

[7] Wu Q, Semenova Y, Wang P and Farrell G 2011 High sensitivity SMS fiber structure based refractometer – analysis and experiment *Opt. Express* **19** 7937

[8] Gao R X, Liu W J, Wang Y Y, Wang Q, Zhao F and Qu S L 2012 Design and fabrication of SMS fiber refractometer for liquid *Sensors Actuators, A Phys.* **179** 5–9

[9] Xia F and Zhao Y 2020 RI sensing system with high sensitivity and large measurement range using a microfiber MZI and a photonic crystal fiber MZI *Meas. J. Int. Meas. Confed.* **156** 107603

[10] Wang T, Liu B, Zhao L, Wu Y, Han Y, Nan T, Wang J, Zheng J and Zhang Y 2021 Enhanced

refractive index and temperature sensor based on balloon-shaped Mach-Zehnder interferometer *Opt. Fiber Technol.* **65** 102615

[11] Yang Y, Wang Y, Jiang J, Zhao Y, He X and Li L 2019 High-sensitive all-fiber fabry-perot interferometer gas refractive index sensor based on lateral offset splicing and Vernier effect *Optik (Stuttg).* **196** 163181

[12] Ferreira M F S, Statkiewicz-Barabach G, Kowal D, Mergo P, Urbanczyk W and Frazão O 2017 Fabry-Perot cavity based on polymer FBG as refractive index sensor *Opt. Commun.* **394** 37–40

[13] Beres C, de Nazaré F V B, de Souza N C C, Miguel M A L and Werneck M M 2011 Tapered plastic optical fiber-based biosensor - Tests and application *Biosens. Bioelectron.* **30** 328–32

[14] Wu Q, Qu Y, Liu J, Yuan J, Wan S P, Wu T, He X D, Liub B, Liuc D, Ma Y, Semenova Y, Wang P, Xin X and Farrell G 2020 Singlemode-Multimode-Singlemode Fiber Structures for Sensing Applications – A Review *IEEE Sens. J.*

[15] De-Jun F, Guan-Xiu L, Xi-Lu L, Ming-Shun J and Qing-Mei S 2014 Refractive index sensor based on plastic optical fiber with tapered structure *Appl. Opt.* **53** 2007

[16] Rahman H A, Harun S W, Yasin M, Phang S W, Damanhuri S S A, Arof H and Ahmad H 2011 Sensors and Actuators A : Physical Tapered plastic multimode fiber sensor for salinity detection *Sensors Actuators A. Phys.* **171** 219–22

[17] Batumalay M, Harun S W, Ahmad F, Nor R M, Zulkepely N R and Ahmad H 2014 Tapered plastic optical fiber coated with graphene for uric acid detection *IEEE Sens. J.* **14** 1704–9

[18] Tiwari D, Mullaney K, Korposh S, James S W, Lee S-W and Tatam R P 2016 Ammonia sensing using lossy mode resonances in a tapered optical fibre coated with porphyrin-incorporated titanium dioxide *Sixth Eur. Work. Opt. Fibre Sensors* **9916** 99161H

[19] Azargoshasb T, Navid H A, Parvizi R and Heidari H 2020 Evanescent wave optical trapping and sensing on polymer optical fibers for ultra-Trace detection of glucose *ACS Omega* **5** 22046–56

[20] Musa B, Kamil Y M, Bakar M H A, Noor A S . and Mahdi M A 2013 Effects of Taper Parameters On Free Spectral Range of Non-Adiabatic Tapered Optical Fibers for Sensing Applications *Microw. Opt. Technol. Lett.* **55** 2562–8

[21] Yang W, Pan R, Yu X, Fan J, Xiong Y and Wu M 2020 A high sensitivity asymmetric double tapered fiber interference sensor *Optik (Stuttg).* **210** 164495

[22] Chen X, Regodon D G, Minguez A M and Horche P R 2013 Mach-Zehnder interferometer based on all-fiber multimode interference device for DPSK signal demodulation *XXVIII Simp. Nac. la Uni fica Int. Radio (URSI 2013)* 1–14

[23] Li L, Leng J, Zhou P and Chen J 2019 Modulation instability induced by intermodal cross-phase modulation in step-index multimode fiber *Appl. Opt.* **58** 4283–7

[24] Walter T F 2012 *Experimental theory and setup*

[25] Korposh S, James S W, Lee S W and Tatam R P 2019 Tapered Optical Fibre Sensors: Current Trends and Future Perspectives *Sensors (Basel).* **19**

[26] Falciai R, Mignani A G, Ciaccheri L and Cosi F Tapered multimode optical fibers for enhanced evanescent-wave absorption spectroscopy of liquids A = aLq S = S =€ Lq **3105**

[27] Lee C, Zeng R, Yang C, Lin C, Ma C, Lee C, Zeng R, Yang C, Lin C, Ma C and Liu W 2020 Tapered Polymer Fiber Inclinometers **12**

[28] Pum J 2019 *A practical guide to validation and verification of analytical methods in the clinical laboratory* vol 90 (Elsevier Inc.)

[29] Molla H A, Bhowmick R, Katarkar A, Chaudhuri K, Gangopadhyay S and Ali M 2015 A novel rhodamine-3,4-dihydro-2H-1,3-benzoxazine conjugate as a highly sensitive and selective chemosensor for Fe3+ ions with cytoplasmic cell imaging possibilities *Anal. Methods* **7** 5149–56

[30] Burgués J, Hernández V, Lilienthal A J and Marco S 2019 Smelling nano aerial vehicle for gas source localization and mapping *Sensors (Switzerland)* **19**

[31] Miller J N and Miller J C 2010 *Statistics and Chemometrics for Analytical Chemistry*

Journal of Physics: Conference Series **2274** (2022) 012008 doi:10.1088/1742-6596/2274/1/012008

Programmable photonic circuits

David Marpaung[1,*]

[1]Nonlinear Nanophotonics Group, University of Twente, the Netherlands.

[*]Corresponding author: david.marpaung@utwente.nl

Abstract. We highlight a recent trend in photonics where integrated photonic chips are designed and operated as programmable devices like in the case of integrated electronics. We discuss the potential application of these programmable photonic circuits in signal processing.

1. Introduction

The maturity of integrated photonic manufacturing has allowed implementation of advanced concepts in integrated optical circuits. Integrated photonic circuits nowadays can have very high number of component counts with each component can be tailored in terms of functionalities. An emerging concept recently embraced by integrated photonics is the concept of programmability [1-3]. In ways that mimic electronics, new generations of photonic circuits are presently composed of basic building blocks that are replicated and interconnected to make a larger circuit that can be programmed to carry out numerous functions.

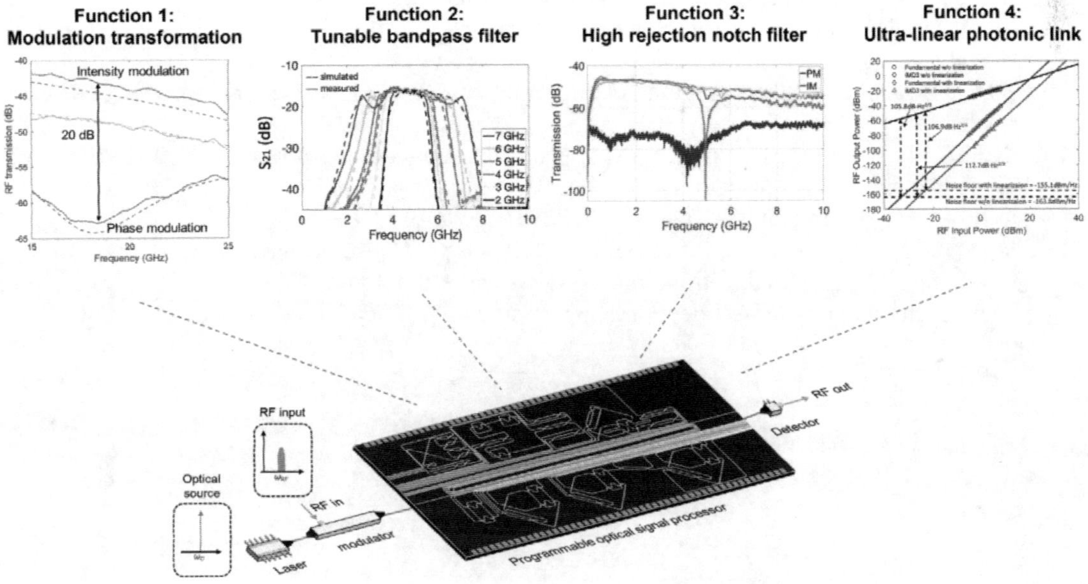

Figure 1. Illustration of a programmable photonic circuit implemented for microwave photonic signal processing.

2. Implementation

In terms of hardware implementation, there are a number of ways to construct a programmable photonic circuit. A more traditional way is to build the circuit based on a mix of dedicated building blocks, such as ring resonators, splitters, combiners, with each of these components capable of reconfiguration [4]. A recent departure from this concept proposed a circuit realized from a mesh of a single building block, which is a tunable beam splitter, that is arranged to allow feedback and feed-forward light propagation [5].

3. Applications

Programmable photonic can find applications in various fields including quantum photonics [1], computing [2], and microwave photonic (MWP) signal processing [3]. For MWP processing in particular, overall system performance including signal-to-noise ratio and linearity is an important factor that currently cannot be satisfied by the mesh-based programmable circuits. For this reason, a new approach in programmable MWP circuit has been proposed. This breakthrough approach relies on two new building blocks with unique programmability, namely an on-chip modulation transformer [6] and a double-injection ring resonator [7]. Using the interconnection of these building blocks an advanced MWP programmable circuit with all-optimized signal processing performance has recently been demonstrated [8].

Acknowledgments

This work is supported by the Dutch NWO Vidi (15702) and Start-Up (740.018.021) grant.

References

[1] Bogaerts W., D. Pérez, J. Capmany, D. A. B. Miller, J. Poon, D. Englund, F. Morichetti, and A. Melloni, 2020 *Nature* **586** 207.

[2] Shastri B.J., A. N. Tait, T. Ferreira de Lima, W. H. P. Pernice, H. Bhaskaran, C. D. Wright, and P. R. Prucnal 2021 *Nature Photonics* **15** 102.

[3] Marpaung D., J. Yao, and J. Capmany 2019 *Nature Photonics* **13** 80.

[4] Daulay O., G. Liu, X. Guo, M. Eijkel, and D. Marpaung, 2021 *Journal of Lightwave Technology* **39** 700.

[5] Pérez D., I. Gasulla, L. Crudgington, D. J. Thomson, A. Z. Khokhar, K. Li, W. Cao, G. Z. Mashanovich, and J. Capmany 2017 *Nature Communications* 8.

[6] Guo X., Y. Liu, T. Yin, B. Morrison, M. Pagani, O. Daulay, W. Bogaerts, B. J. Eggleton, A. Casas-Bedoya, and D. Marpaung, 2021 *APL Photonics* **6** 36106.

[7] Cohen R. A., O. Amrani, and S. Rushcin 2018 *Nat. Photon.* **12** 706.

[8] Daulay O., G. Liu, R. Botter, M. Hoekman, E. Klein, C. Roeloffzen, J. Capmany, and D. Marpaung "Programmable Integrated Microwave Photonic Filter using a Modulation Transformer and a Double-Injection Ring Resonator", *ECOC 2021* (accepted as high-scored paper)

Journal of Physics: Conference Series **2274** (2022) 012009 doi:10.1088/1742-6596/2274/1/012009

The development of the real-time Michelson interferometer for the measurement of thermal expansion coefficients of dental composites nanofiller

R. Apsari[1], Halimatus Sya'dyyah[1], Kristia Ningsih[1,2], and Y. G. Y. Yhuwana[1]

[1]Department of Physics, Faculty of Science and Technology, Universitas Airlangga, 60115, Indonesia
[2] Optoelectronics Research Centre, The University of Southampton, Highfield Campus, Southampton SO17 1BJ, UK

Corresponding author: retna-a@fst.unair.ac.id, kn1n17@soton.ac.uk

Abstract. This study aims to design a thermal time-based expansion coefficient sensor system using a nickel interferometer that uses a 6 Watts heater to produce a similar oral cavity temperature of 30°C to 60°C. The light source used was the He-Ne Laser, with a wavelength of 632.8 nm. Hollow cylindrical samples with an outer diameter of 1.1 cm, an inner diameter of 1 cm, and a height of 1.5 cm were placed in the sample box behind one of the Michelson Interferometer mirrors. The interference pattern (fringes) formed was recorded using a webcam and counted using the principle of motion detection in the Delphi program. The temperature sensor used was LM 35 with LM 358 amplifier. The resulting voltage output was converted to a temperature value with Arduino. The data were then displayed on a PC with the Delphi interface. The purposely written Delphi program has four functions: recording and counting fringes, displaying temperature, and calculating the thermal expansion coefficient of acrylic resin and nanofiller composite. The generated data in the study included the heating time for 25,421 seconds to produce a temperature of 60°C, delay of (0,802 ± 0,006) second with the sensor performance of 99,76%, and software performance of 98,18%. In three data collections, the average thermal expansion coefficient of resin acrylic was (111.37 ± 10.93) .10-6 / ° C with an error percentage of 23.74% from the literature, while for the composite nanofiller samples, we obtained (49.6 ± 0.95) .10-6 / ° C with an error percentage of 6.9% from previous studies.

1. Introduction

Based on the World Health Organization data, *WHO Oral Health Media Center*, in 2012, dental cavities are recently still believed as one of the most frequent diseases as many as 60-90% experienced by children and adults. This issue can cause severe health problems penetrating the body's other organs. Therefore, dental filling is one of the efforts to overcome tooth cavities that effectively reduce pain. However, the dental fillings materials must have a specific endurance to various treatments, one of which is thermal changes. Thermal resistance is related to tooth sensitivity level. Nevertheless, the greater of the thermal expansion coefficient mismatch, the greater possibility of fluid percolation below the margin, which can cause caries [1].

Furthermore, restoration teeth undergo more significant expansion or contraction than natural teeth during temperature changes. Thus, restoration may leak or fall apart from the tooth [2]. Therefore, it is crucial to understand the value of the thermal expansion coefficient in dental fillings.

Along with the increasing needs of dental fillings, new fabrication techniques of dental filling materials are continuously being developed. The selection of dental fillings materials is based on several considerations of properties: biocompatibility, the nature of physical chemistry, the characteristics of the handling, the aesthetics, and the economic factor [2].

The detection methods of thermal expansion coefficient that have been used are Differential Thermal Analysis (DTA) and Differential Scanning Calorimetry (DSC). However, those available methods are rather complicated and expensive [3]. Many measurements of the thermal expansion coefficient of dental fillings material encourage the needs to invent alternative methods. One of the alternative prospective methods is the use of optical methods. The advantages of optical methods are the high precision, non-invasive, uses the non-destructive source.

Consequently, it has minimal side effects and it can be observed by visual [4]. In Scholl et al., 2009 [5] study stated that the Michelson Interferometer was successfully used to measure the thermal expansion coefficient. The interferometry method can also be implemented to calculate the thermal expansion coefficient of ZnSe crystalline thin material [6].

The research conducted by Ariyanti in 2008 [7] implemented the interference and coherence concept at the Laboratory of Physical Optics and Laser Department of Physics, Airlangga University. It utilized the real-time Michelson Interferometer to detect the deformation of teeth due to temperature changes. This study's weakness is the number of delay time reaches (1.8 ± 0.7) seconds between the temperature displayed in the digital thermometer (calibrator) and the temperature resulting from the circuit system of temperature, Arduino, and Delphi. Another weakness is the visual observation of the tassel fringes.

Further research conducted by Atmawati in 2012 [8] was optimizing the microcontroller AT Mega 8535 and temperature sensor LM 35 and the counting of fringes using the principle of motion detection at Delphi in real-time. The result obtained that the delay time amounting to (1.1 ± 0.1) seconds. The weakness of the study is the process of sample heating by using a high-power solder, so the materials undergo a quick increase in temperature. The additional weakness is that the principal terms related to the sample thinness cause the samples preparation more complicated to perform.

The Mach Zender Interferometer modulation phase in Atmawati's study 2012 [8] was located on a particular sample placed in one of the interferometer arms. This method caused the limited number of the chosen samples, which were the thin and transparent samples allowing the light transmission source. The research method used a sample in a tube with an inner diameter of 1 cm, an outer diameter of 1.1 cm, and a height of 1,5 cm. The sample was placed in the nickel winding in the back of one of the mirrors. The length changes due to heating would be detected by shifting the mirror. Then, it would be observed from the fringe patterns on the screen. In addition to the research from Ariyanti [7] and Atmawati [8], Maiorov [9] also carried out research developments in the dental field using an Interferometer. Thus, it generated interesting measurements for the dentistry developments. Therefore, this research was expected to give contribution to develop or apply as non-invasive methods.

The dental fillings materials used as samples in this research were resin acrylic and composite nanofiller. The study by Marquis et al., 2010 [10] stated that the material of composite nanofiller was used since it was the newest material of dental fillings with the structure of the filler material that has nanometer size; thus, it was expected to have a better resistance against the temperature.

The samples used in this study were treated with temperature. The temperature variations permitted to be detected the deformation on the teeth range from 30° to 60° C following the temperature in the mouth's environment [11]. The previous research used solder as the heating element. The amount of power in the solder made the samples experience more rapid expansion. Hence, it was necessary to design a lower power heater following the heating characteristics in the mouth cavity.

In this study, we employed a He-Ne laser light source with a wavelength of 632.8 nm. The interference pattern (fringes) was recorded continuously [6]. The analysis of recorded results of the fringes' pattern used the principle of motion detection. The chosen method of the principle of motion detection has been successfully utilized to calculate fringes generated by the Michelson interferometer

[8].

An Arduino, temperature sensor circuitry, power supply, nickel winding as heating, mechanics box, a laptop, and a webcam were employed to achieve the real-time conditions on the study. LM 35 temperature sensor was used to capture the temperature changes. Arduino Uno was used to changing the temperature sensor's output from analog to digital, and then it was sent to a laptop with serial communication to be displayed using the Delphi program. If the sample temperature reached 30°C, Delphi would automatically record and count the fringes that appeared. After the temperature of the sample showed 60ºC, Delphi stopped the recording. The Delphi software was used to calculate the number of fringes patterns for analysing the thermal expansion coefficient on the dental filling materials of resin acrylic and composite nanofiller. Furthermore, a comparison was performed with the research results from Atmawati, 2012 [8] and the literature to determine the proposed set-up performance.

The study's achievement was expected to be an initial step to apply the Michelson Interferometer based on laser and microcontroller in real-time to characterize dental fillings material and is a non-invasive treatment in dentistry. Further development can detect dental fillings material suitable to use and resist the thermal change [10].

2. Experiment Set-Up

The first stage conducted in the research was to design the hardware, presented in Figure 1. Then, the next stage was to design the software for computing the research purpose. The software was built with Delphi 7. It consisted of recording and counting the fringes, the program of temperature viewer, and the program of calculating the thermal expansion coefficient.

After the hardware and software design stage was completed, the next stage was the calibration stage. To calibrate the hardware, we calculated the resulting heating time, which ranges from 30°C to 60°C following the cavity temperature of the mouth. The calibration was also performed to determine the voltage conversion value to the temperature by noting the sensor's voltage output value to the value displayed on the digital thermometer. The software was also calibrated to measure the wavelength of the He-Ne Laser light source.

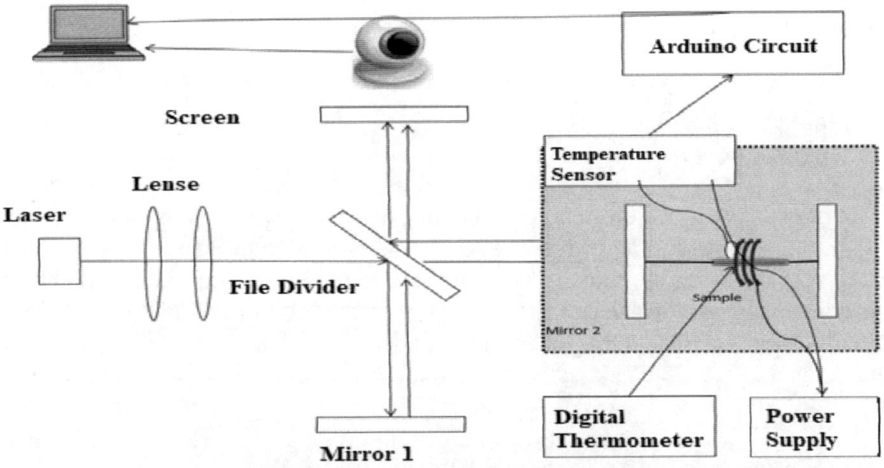

Figure 1. Research set-up scheme

The process of collecting the data was conducted three times for each sample. The samples mold was a 3 cc spad with an outer diameter of 1 cm. Before being molded, the spad outside surface was firstly smeared with Vaseline to easily facilitate the sample's release when the sample has dried. The sample of resin acrylic was in the form of polymer powder and monomer solution. The powder and solution were weighed in a ratio of 1:1 and then stirred until homogeneous. A composite nanofiller sample, a gel, was ready to be used. Before it was dried, the sample was attached to the surface of the spad. Resin acrylic was used as a self-heating to dry out in a temperature chamber for 5 minutes. The Halogen lamp was used for drying of composite nanofiller sample. The dried samples were mashed up to an outer diameter of 1.1 cm. The height of the sample was cut to 1.5 cm. The maximum thickness of the dental filling materials was 2 mm [12]. In this study, the samples were molded with a thickness of 0.5 mm. The consistency was reasonably small, allowing the fast diffusion of heat.

3. Result and Discussion

Hardware in the study consisted of a power supply circuit, heater and a sample box, temperature sensor, and Arduino Uno. It included three functions: heat the sample, measure the temperature, and send the data to a laptop to be displayed on the Delphi program. The overall hardware is presented in Figure 2

Figure 2 Overall hardware circuit

Description:
A: The circuit of the power supply
B: The circuit of the temperature sensor
C: Arduino Uno Board
D: Nickel winding for heating
E : LM 35 temperature sensor

The system's program was built to record and count fringes, display temperature, and calculate the coefficient of thermal expansion. The program's display is presented in Figure 3.

The first step was inputting the images from the source recorder used (webcam). The device used was an external webcam with A4 Tech USB PC Camera. The program also had device settings properties that adjusted the device's brightness, hue, and white balance. Therefore, the expected result would be captured as a video. The next step was to choose the resolution of the video, as presented in Figure 4.

In this study, the higher resolution would gain the better quality of the video. The proper resolution was the highest resolution of 640 x 480 pixels without any compression.

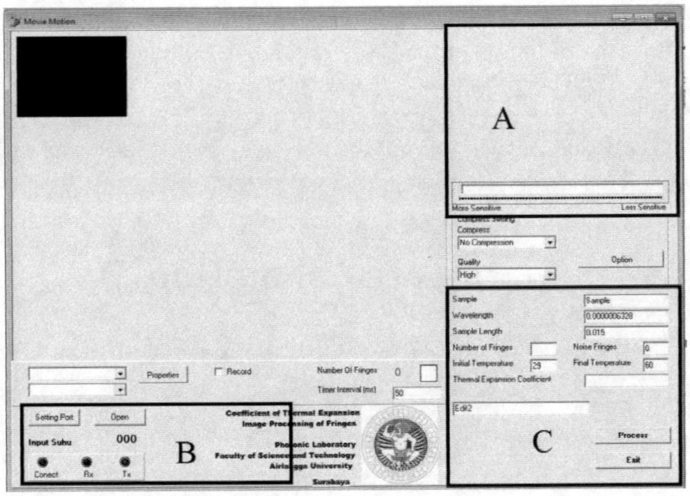

Figure 3 The display of the entire program

On the counting program of thermal expansion coefficient, the parameters included the wavelength of He-Ne laser, the number of fringes including the system's noise, the length of samples, and the range of temperature. The optimal temperature range was 30°-60°C as the range temperature of the oral area (Kishen et al., 2001). The length of the sample used was 1.5 cm with a thick sample of 0.5 mm, while the maximum thickness of the dental fillings material was 2 mm [12].

The noise data fringes were obtained from the number of fringes counting. The counting started when the heater was turned on without a sample. The result of fringes counting due to heat treatment of the sample was the number of fringes at the time of treatment subtracted by the number of noise fringes.

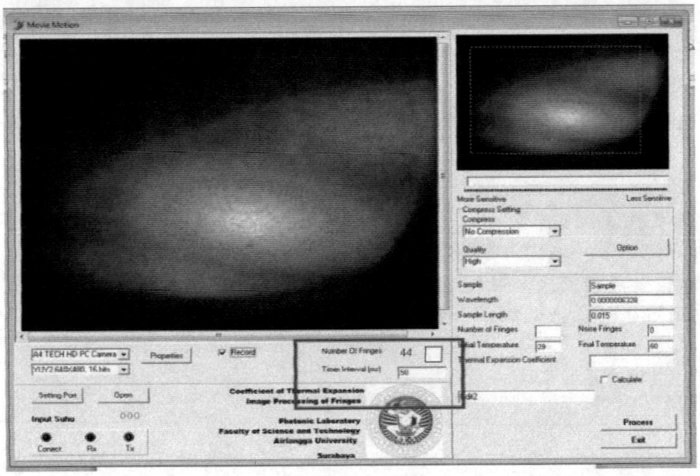

Figure 4 The fringes recording and counting program.

By using the data calibration of the output voltage value, we could convert the voltage into temperature by reversing the relationship between voltage output and temperature of the linear regression relation in Fig. 5. The output voltage was *x*, and the displayed temperature was *y*; the relationship between voltage and temperature was obtained, $y = 25.711x - 6.1739$ with a coefficient of determination of 0.99. The value conversion of hardware was included in the Arduino program as the voltage conversion to the temperature form.

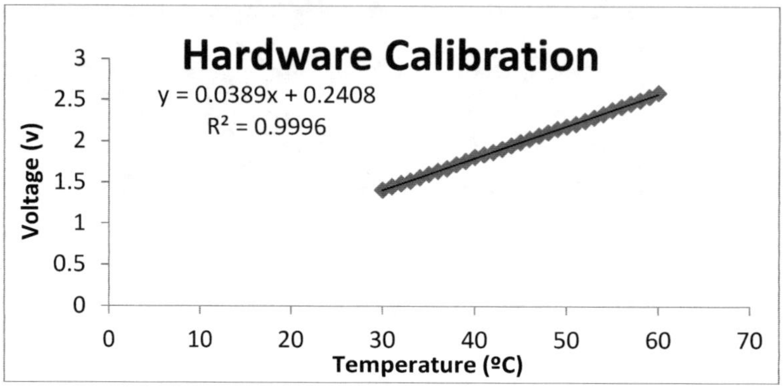

Figure 5 The graph of the relationship of sensor output voltage to the temperature

Software calibration was done by observing the number of fringe pulse counting when the micrometer was shifted in the range of a particular length which as 10μm. There was a camera connected to the program, which replaced the function of the eyes. Furthermore, the webcam recorded and counted the number of fringes simultaneously displayed on the screen during the shifting.

The wavelength of the He-Ne laser source generated by the software calculation was 632.8 nm. From the test results of the He-Ne laser, the software performance results were 98.18%, with a sensitivity value of 25 using a 50 ms time interval. From the data performance software, it concluded that the system could not capture the number of counted fringes although it had used the smallest sensitivity value and interval timer. The performance of the software could be upgraded by replacing the recording instrument with a much higher frame rate, which is more than 30 frames/second.

The built heater was a nickel winding powered by an electric current from a voltage source of 8.4 volts. Figure 6 shows a graph of the relationship between temperature rise and the needed time. The heat from the temperature 30°C - 60°C requires a time of 25 seconds. The length of the time is within the range of the length of chewing for humans, which is 10-30 seconds. The value of the regression graph approaches one, which indicates that the heating characteristics in the windings of nickel are linear to heating time.

Journal of Physics: Conference Series **2274** (2022) 012009 doi:10.1088/1742-6596/2274/1/012009

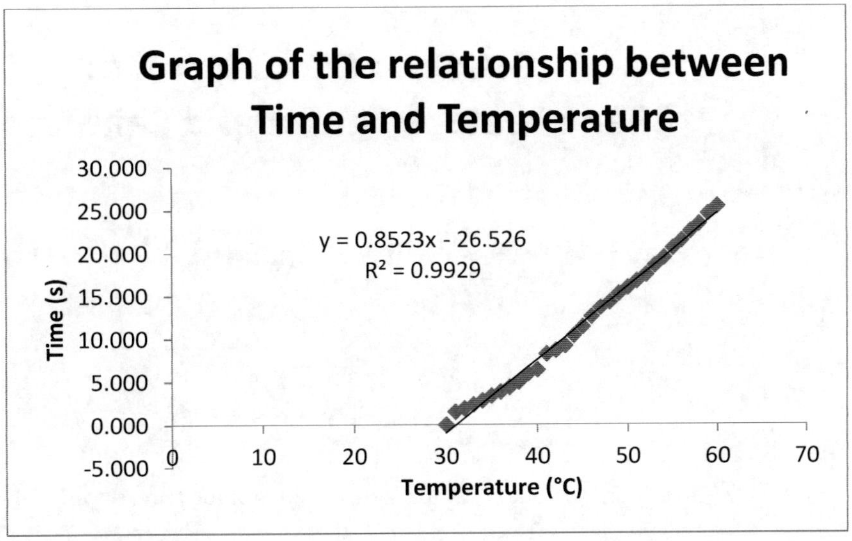

Figure 6 The graph of the relationship between time and temperature

The delayed testing was used to determine the real-time condition, achieved by the system. It was obtained from the measurement of when the temperature appeared in the program and when the temperature was displayed on a digital thermometer (calibrator). From the two generated timings, Δt was evaluated as the delay. Based on the calculation, the acquired delay was 0.006 seconds.

The received LM35 temperatures were the average between air temperature and surface temperature contacted with heaters. For LM 35 TO 92 plastic package, with lead copper as track of heat carrier to the device, it made the temperature closer to the environment temperature than the contact surface (datasheet LM 3). To minimize this issue, cover the lead of copper and all cables with materials. However, it had to ensure that the whole system was at equal temperature to the contact surface temperature and the air temperature did not affect the readable temperature. The relatively small heaters' power created a more prolonged time to heat the temperature of the sensor's environment, thus forming the delay of measurement between calibration and the sensor temperature.

Sensor performance testing aims to determine the sensor's accuracy compared to a digital thermometer as a temperature calibrator. The sensor and calibrator comparison chart is presented in Figure 7. Based on Figure 7, it discovers the coefficient of determination is 0,9976 so the sensor performance compared with a digital thermometer is at 99.76%.

LM35 has a tolerance of 1.5°C at temperatures above 25°C and 2°C for minimum temperatures (0°C) or maximum temperatures (100°C). There are five differences in the readable temperature on the sensor and the calibrator, four of which have a difference of 1°C, which means still within the sensor's temperature reading tolerance. Meanwhile, one datum at a temperature of 35°C has a difference of 2°C, an error measurement system sensor.

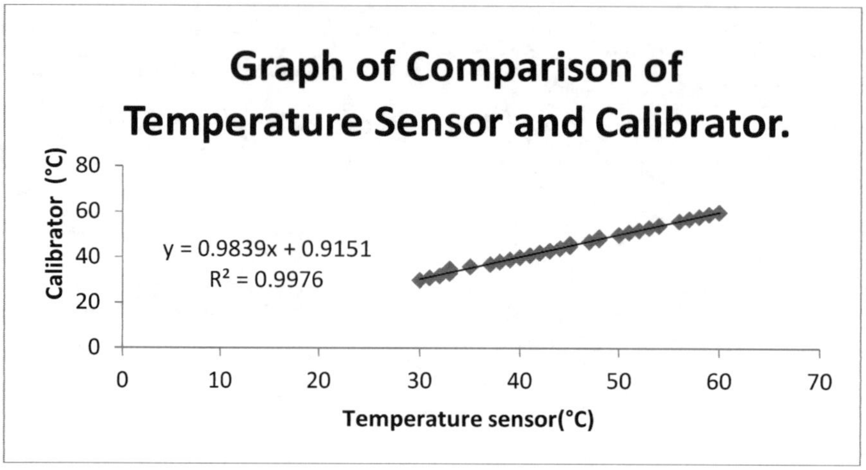

Figure 7 The graph of comparison of temperature sensor and calibrator.

The data was collected using the built real-time Michelson interferometer to count noise due to the environment and the system as it is delivered in Table 1. The number of fringes counted at the Michelson Interferometer System was run without extensive samples. Executing data collection at night could minimize this. The system would result in less precise measurements when used to measure the thermal expansion coefficient, resulting in fewer fringes than its noise. The prepared resin acrylic and composite nanofiller samples were inserted into the box of samples. The data collection of the samples was taken three times. Each sample was performed on a different day.

Table 1. The number of fringes counted by the system without a sample (Background)

Temperature (°C)	Number of Fringe Counting	
	Data Collection 1	Data Collection 2
30	12	3
35	36	8
40	48	14
45	70	23
50	84	39
55	98	53
60	112	72

Resin acrylic sample measurements were performed on the first day. Before measuring the sample three times, noise system measurement was conducted first. The data collection results of resin acrylic samples are presented in Table 2.

After calculating the fringes on a system without samples for initial calibration, the calculations were carried out using acrylic resin samples. The value of the coefficient of thermal expansion of resin acrylic, according to Spiller (2011), is $90.\,10^{-6}/$ °C. Based on the calculation results, the value of the coefficient of the thermal expansion of resin acrylic was $(\,99{,}1 \pm 5{,}12).\,10^{-6}/$ °C, $(\,116{,}0 \pm 5{,}1).10^{-6}/$ °C, and $(119{,}0 \pm 18{,}1).\,10^{-6}/$ °C respectively. The mean value of the coefficient of thermal expansion of resin acrylic of the third session of data collection was $(111{,}37 \pm 10{,}93).\,10^{-6}/$ °C with a percentage error of 23.74% to the literature (Spiller, 2011).

The relation between the observed fringe count and the thermal expansion coefficient of the dental filling is shown in equation 1. The greater fringes number leads to the greater dental filling thermal coefficient and vice versa. This is in accordance with this study's result figured in Table 1, Table 2, and Table 3.

$$\alpha = \frac{n\lambda}{2L_o\Delta T} \quad (1)$$

where α = Coefficient of thermal expansion, n = Number of changes in dark and light pattern, L_o = Thickness of the initial sample, ΔT= Temperature change.

Table 2. The results of data collection of resin acrylic samples

Temperature (°C)	The number of Fringes Counting			
	Noise	Data Collection 1	Data Collection 2	Data Collection 3
30	12	38	32	18
35	36	86	70	74
40	48	132	112	112
45	70	172	158	160
50	84	210	200	208
55	98	252	248	248
60	102	274	294	288

The data collection on a sample of composite nanofiller is presented in Table 3. The value of the coefficient of thermal expansion based on the literature is $(53{,}4 \pm 0{,}6).10^{-6}/$ °C (Park et. al. 2011). The results of this study were $(50{,}8 \pm 2{,}3).10^{-6}/$ °C, $(49{,}7 \pm 2{,}7).10^{-6}/$ °C, and $(48{,}5 \pm 2{,}9).10^{-6}/$ °C respectively. The obtained value of the thermal expansion coefficient from the calculation is $(49{,}6 \pm 0{,}95).10^{-6}/$ °C, with a percentage error of 6.9% to the research of Park et.al. (2011).

Table 3. The results of data collection of composite nanofiller samples

Temperature (°C)	The number of fringe counting			
	Noise	Data Collection 1	Data Collection 2	Data Collection 3
30	3	11	12	12
35	8	21	24	24
40	14	40	48	42
45	23	62	64	62
50	39	88	88	86
55	53	118	120	114
60	72	150	152	152

The correlation between the results of fringe count and thermal expansion coefficient is to determine the error factors during treatment. Moreover, the high percentage of error is alleged caused by several factors, including:

- The dough resin acrylic samples are less homogeneous.
- The quality of the sample experienced a decline due to the thermal treatment, so there is a quite large difference in each data collection process, as shown in Table 2 and Table 3.

- There is a 0.8 seconds delay allowing the loss of information. This is due to the rising heat temperature; meanwhile, the sensor is late in reading the temperature value.
- The ability of camera to record the video is restricted to 30 frames/second. Thus, there is a possibility that some fringes are not counted.
- The differences in material composition in the literature and research. Thus, the research sample needs to be tested by TMA (Thermomechanical Analysis) as implemented in Park et al.'s (2011) research.

According to Spiller, 2011 [13], dental dentin has a thermal expansion coefficient of $9.10^{-6}/$ °C. The bonding of dental fillings materials on teeth can be changed if the dental fillings materials and teeth have differences in coefficient of expansion of thermal [12]. It is because the tooth is a very good isothermal [14]. Shrinkage of polymerization can cause gaps between teeth and filling material, resulting in leakage [15-16].

In this research, the designed system of real-time Michelson interferometer could be used to measure the thermal expansion coefficient of dental filling materials. The design minimized the delay by 27% compared to Atmawati's study 2012 [8] by replacing the used microcontroller. Measurement of thermal expansion coefficient also produced an error percentage that was smaller than Atmawati's study 2012 [8], with the replacement of the transmission design to reflection. The heating system had reasonably low power. The heating time also corresponded to the time of food chewing in the oral cavity. Therefore, it concluded that the heater was built acceptable to the oral cavity temperature.

Based on the study's result, this system can be continuously applied to shifting micro-materials order, which is sensitive to heat changing. The built-in real-time system is also able to have further development by replacing the temperature sensor or coating the copper on the cable to prevent outside temperatures. The heating design can also be continually improved by replacing the nickel with a biocompatible material suitable with the conditions of the cavity mouth, not only the temperature. The homogeneity and purity of the samples were firstly tested because the sample composition determines the significant coefficient of thermal expansion. The system's stability is relatively low, as it is represented in the collection of background counting in Table 2 and 3, hence its application is limited to the state of minimal noise. Manufacturing the optical bench design can fix this condition to reduce the environmental vibrations. The high sensitivity level of the Michelson Interferometer also leads to the occurrence of the changing fringes pulse in the air density on the heat condition. Thus, it requires the insulation of the laser beam by using a fiber interferometer.

4. Conclusions

Based on the data analysis, the research obtained that the constructed system of real-time Michelson Interferometer can be used to measure the thermal expansion coefficient of tube-shaped material in dental fillings. The average result of three times calculation was $(111{,}37 \pm 10{,}93).10^{-6}/$ °C, with an error percentage of 23.74% in resin acrylic samples. Meanwhile, in composite nanofiller samples, the average value of the thermal expansion coefficient of three times data collection was $(49{,}6 \pm 0{,}95).10^{-6}/$ °C, with an error percentage of 6.9% compared to previous studies with the delay system of $(0{,}802 \pm 0{,}006)$ second, the sensor performance of 99.76%, and the software performance of 98.18%. Furthermore, this research is expected to give a non-invasive technology-based development, especially in the dentistry field.

References

[1] Spiller M.S. 2011. *Dental Composites: A Comprehensive Review*. Academy- Dental Learning and OSHA training. Minnesota.
[2] Anusavice K. J. 1996. *Phillips' Science of Dental Materials, 10th Edition*. WB Saunders, Philadelphia.

[3] Setyabudi. 2010. *Thermal Analysis*. Universitas Padjajaran. Bandung.
[4] Apsari R. 2007. Development of Interferometer Based on Electronic Speckle Pattern Interferometry for in vitro Temperature Deformation Analysis in Teeth. UNAIR. Surabaya.
[5] Scholl R. and Liby B. W., "Using a Michelson Interferometer to Measure Coefficient of Thermal Expansion of Copper," *Phys. Teach.*, vol. 47, no. 5, pp. 306–308, 2009, doi: 10.1119/1.3116844.
[6] Su C. H., Feth S., and Lehoczky S. L., "Thermal expansion coefficient of ZnSe crystal between 17 and 1080 °C by interferometry," *Mater. Lett.*, vol. 63, no. 17, pp. 1475–1477, 2009, doi: 10.1016/j.matlet.2009.03.050.
[7] Ariyanti R. 2008. Development of Michelson Real Time Interferometer for Detecting Temperature Deformation in Teeth. Thesis. Universitas Airlangga. Surabaya.
[8] Atmawati E.U. 2012. Michelson Real Time Interferometer Optimization for Detection of Thermal Expansion Coefficient of Nanofiller Composites. Thesis. Universitas Airlangga. Surabaya.
[9] Maiorov E.E., Shalamai L.I., Chernyak T.A., Dagaev A.V., Khokhlova M.V., Guliev R.B.. 2020. In Vitro Investigation of Dental Enamel by Shift Interferometry.Biomedical Engineering. Springer Science+Business Media. LLC.
[10] Marquis D.M., Guiilaume E., Chivas-Joly C. 2010. *Properties of Nanofiller in Polymer*. Intechopen
[11] Kishen A., Murukeshan V. M., Krishnakumar V., and Asundi A., "Analysis on the nature of thermally induced deformation in human dentine by electronic speckle pattern interferometry (ESPI)," *J. Dent.*, vol. 29, no. 8, pp. 531–537, 2001, doi: 10.1016/S0300-5712(01)00048-3.
[12] Philips R.W. 1982. *Skinners Science of Dental Materials, 8th Edition*. WB Saunders Co, Philadelphia. Tokyo
[13] Spiller M.S. 2011. *Dental Composites: A Comprehensive Review*. Academy- Dental Learning and OSHA training. Minnesota.
[14] Firdausy K. and Yudhana A., "Webcam for Monitoring System Using Motion Detection Method" vol. 2008, no. Snati, pp. 2–7. National Seminar of Information Technology Application (SNATI 2008).
[15] Farahanny, Wandania. 2009. *Posterior Composite Resin Restoration Techniques-Class II Mod*. Universitas Sumatra Utara. Medan.
[16] Noort VR. 2002. *Introduction to dental materials*. 2nd ed. Mosley; 2002. p. 81–93. London

Characterization and driving of polymer-dispersed liquid-crystal sheet for smarthome applications

Henri P. Uranus[1,*]

[1]Department of Electrical Engineering, Universitas Pelita Harapan, Jl. M. H. Thamrin Blvd. 1100, Tangerang 15811, Indonesia

[*]Corresponding author: henri.uranus@uph.edu, h.p.uranus@alumnus.utwente.nl

Abstract. A commercial polymer-dispersed liquid-crystal (PDLC) sheet sample was characterized for its basic optical and electrical properties under DC and AC excitation. The results give driving voltage for 90% of transmittance for the PDLC of 10 V for DC excitation, and 9.5 V_{rms} for 50 Hz AC excitation. The measured contrast ratio of the PDLC is 2.55 and 2.47 under DC and AC excitation, respectively, which is suitable for smart window in a smarthome application. While it has rather similar optical properties under DC and AC excitation, it has significantly different electrical properties for DC and AC excitation. The PDLC exhibits a capacitive property which has capacitance per unit area of 18.5 $\mu F/m^2$, indicating that DC excitation will give benefit compared to AC excitation which will scale with area of the PDLC. Based on these results, a DC driving circuit employing boost converter was designed to make a microcontroller-compatible driving circuit. The system was then integrated into a miniature smarthome demonstrator model exhibiting contrast ratio of 2.05 under switching from 0 to 100% of duty cycle of pulse width modulation (PWM) control signal from the microcontroller executed through voice command in an internet of things (IoT) platform.

1. Introduction

One of technologies developed in the Industry 4.0 era is the internet of things (IoT) [1], which enables machine-machine and man-machine interaction through the internet. One of implementation of IoT is the smarthome [2], in which smart devices embedded within home components or home appliances can be read or controlled through the internet. One of components within a smarthome is the smart window, which is a window that can be made transparent or white translucent by command from the smarthome control circuit. The core of typical smart window is the polymer-dispersed liquid-crystal (PDLC) [3] sheet, which is an electrochromic sheet with microdroplets of liquid crystal dispersed within a solid polymer sheet matrix which can be laminated onto a normal glass window.

In most smart window implementations, AC driving voltage was used for PDLC, since AC source is normally easily accessible and transformed within the installed location. However, DC voltage also works for PDLC with the cost of an extra DC/DC boost circuit or simply a rectifier circuit. How does PDLC differ under DC and AC excitation voltage in its optical and electrical properties is interesting to be investigated. In this work, a sample of commercial rectangular PDLC sheet was characterized under DC and AC excitation voltage for its basic optical and electrical properties and their comparison are discussed.

Based on the characterization results, an electronic driving circuit compatible for supply voltage of DC 5 V normally used in microcontroller system was designed and tested to excite the PDLC. The

PDLC sheet and the driving circuit, were then combined with other sensors and actuators to form a miniature smarthome demonstration unit [4].

2. The PDLC Sample

The working concept of the PDLC is illustrated in Fig. 1, where microdroplets of liquid crystal (LC) is dispersed within solid polymer matrix of refractive index n_p sandwiched between two parallel transparent electrodes. Without excitation voltage, the orientation of the LC molecules are in equilibrium that is not aligned to the light direction, leading to effective refractive index n_{eff} of between its ordinary (n_o) and extraordinary (n_e) index, which does not match to n_p [5]. Hence, transmitted light will be scattered by the microdroplets, and the PDLC sheet looks white translucent, leading to a low transmission. When sufficient electrical voltage exceeding the threshold voltage (V_{th}) is applied to the electrodes, the electric field is large enough to produce electric torque that overcomes the restoring elastic torque of the LC molecules [6]. Hence, the LC molecules will be aligned to the electric field direction, which changes its effective refractive index n_{eff} to be n_o as light travelling in the direction of its optical axis. The PDLC is designed to have n_o close to n_p, hence the light will pass the sheet with high transmission.

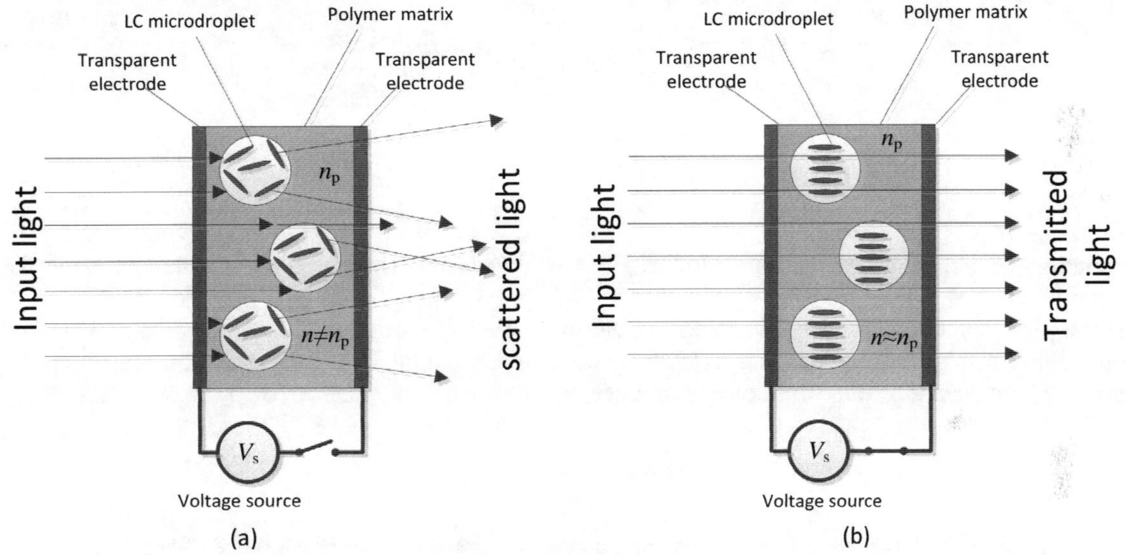

Figure 1. The working concept of PDLC: a. without excitation, b. excited with $V_s > V_{th}$.

The PDLC used in this work is a commercial sample purchased from online marketplace [7] of rectangular sheet with size of 155 mm x 100 mm and thickness of 370 μm. The sheet is claimed by the factory to have 200 ms response time, power consumption of < 5 W/m^2, on transmission > 78%, off transmission < 1%, viewing angle of 150 °, and operation ambient temperature from -30 °C to 60 °C.

3. Measurements and Discussions

The PDLC sheet's optical and electrical characteristics were measured using following procedures.

3.1. Transmittance measurement under DC excitation

The transmittance of the PDLC sheet was measured using set-up shown in Fig. 2. A 20 W/12 V halogen lamp with collimating lens was used as light source. The PDLC sheet is placed 5 cm away from the lamp onto a PMMA plate of Marga Cipta brand with thickness of 2 mm. This PMMA plate will be used as window in the smarthome application that will be discussed later. A silicon sensor head S120C of Thorlabs' optical power meter (OPM) PM100D was placed 5 cm away from the PDLC sheet. For this measurement, the power meter was calibrated for wavelength λ = 635 nm. We should note that, since transmittance is a rational quantity, measurement results will be self-calibrated for white light source from the halogen lamp. The PDLC sheet was excited by a variable DC power supply GPS-4303. All apparatus were mounted on an optical laboratory rail.

In this measurement, the transmittance was measured for various value of DC excitation voltage. Firstly, the power meter readout P_0 for similar set-up, but without the PDLC was used as reference. Then transmittance of subsequent measurements with varied excitation voltage can be calculated using $T = P/P_0$, or $T_{dB} = 10\log(P/P_0)$ with P is the power meter readout for respective excitation voltage.

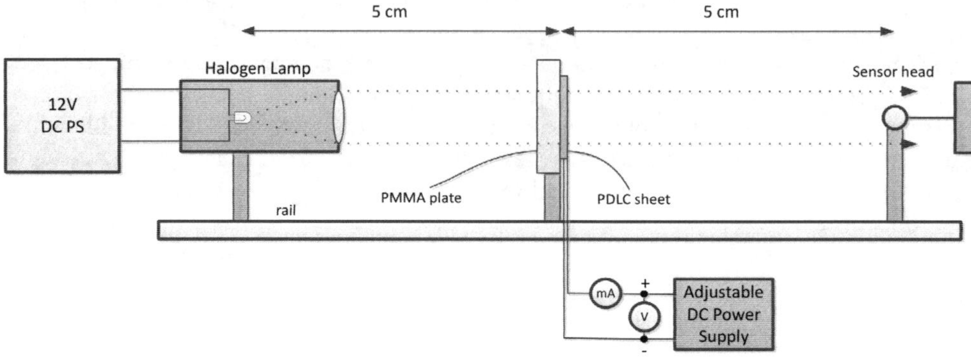

Figure 2. Measurement set-up for PDLC under DC excitation.

The results are shown in Fig. 3, from which we can reveal the V_{90}, the driving voltage (i.e. excitation voltage for 90% of transmittance) of 10 V, and V_{99} (i.e. excitation voltage for 99% of transmittance) of 23.95 V. The figure also shows that without being excited, the PDLC already exhibits 39% of transmittance due to the scattering, hence the term of threshold voltage V_{th} is not applicable for this sample. We also tried to reverse the excitation voltage source polarity during the experiments which gave the similar transmittance properties.

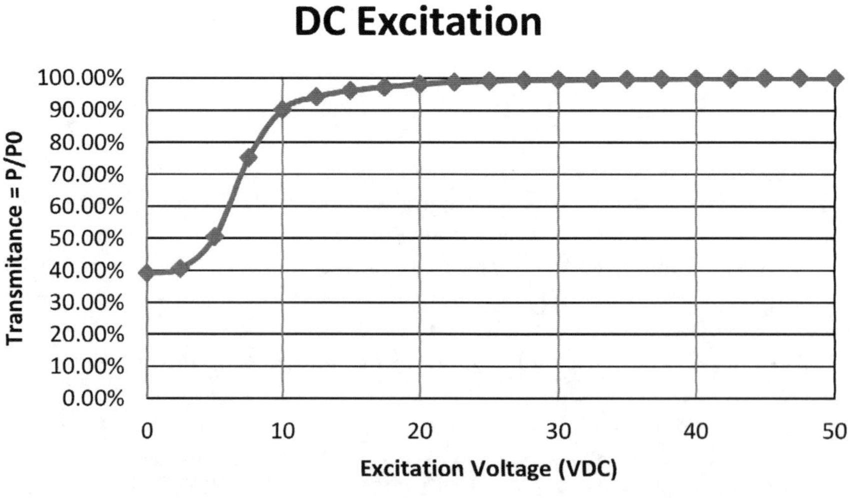

Figure 3. Transmittance of the PDLC under DC excitation.

Fig. 4 shows the contrast ratio = T_{on}/T_{off}, where T_{on} is the transmittance under excitation (on or transparent state), while T_{off} is transmittance under no excitation (off or white-translucent state). The results show that the maximum contrast ratio for this device is 2.55.

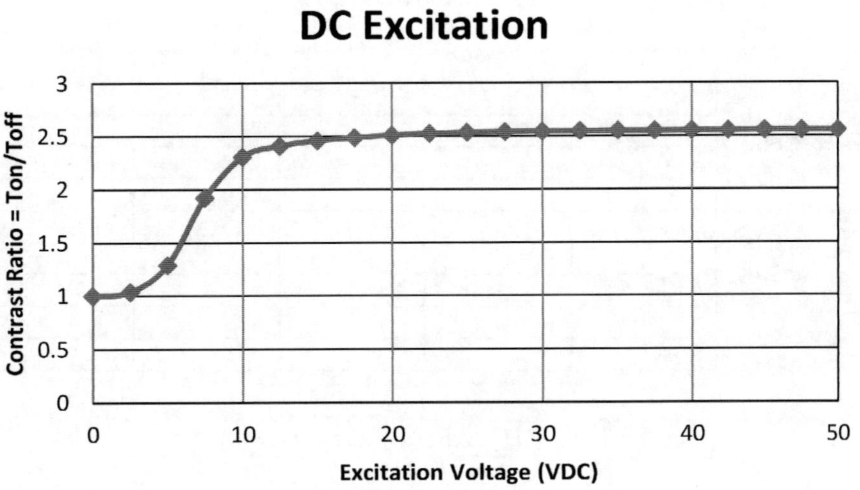

Figure 4. Contrast ratio of the PDLC under DC excitation.

3.2. Transmittance measurement under AC excitation

The measurement set-up under AC excitation is shown in Fig.5. In this set-up, a 32 V_{rms} step-down transformer was fed from the 220 V_{rms} 50 Hz power line, and a potentiometer of 1 kOhm was used to vary the excitation voltage.

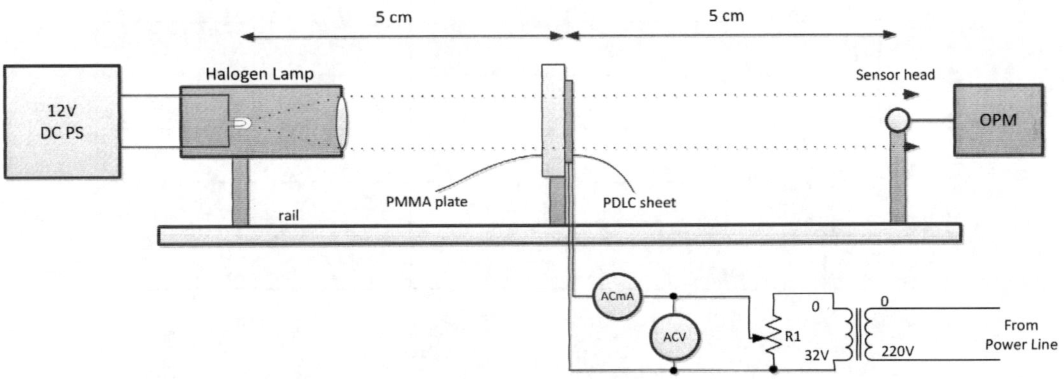

Figure 5. Measurement set-up for PDLC under 50 Hz AC excitation.

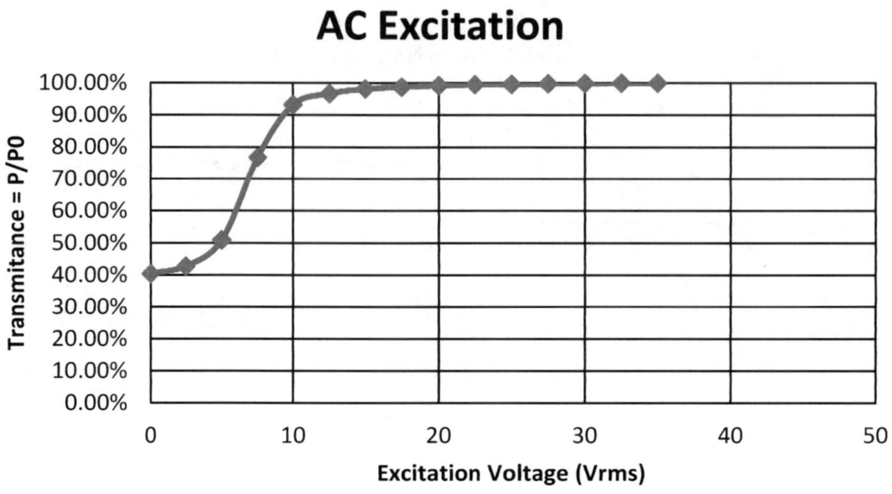

Figure 6. Transmittance of the PDLC under AC 50 Hz excitation.

The results of transmittance measurement under AC excitation are shown in Fig. 6 which shows that V_{90} is 9.5 V_{rms}, while V_{99} is 18.63 V_{rms}. Fig. 7 shows its contrast ratio which gives maximum contrast ratio of 2.47.

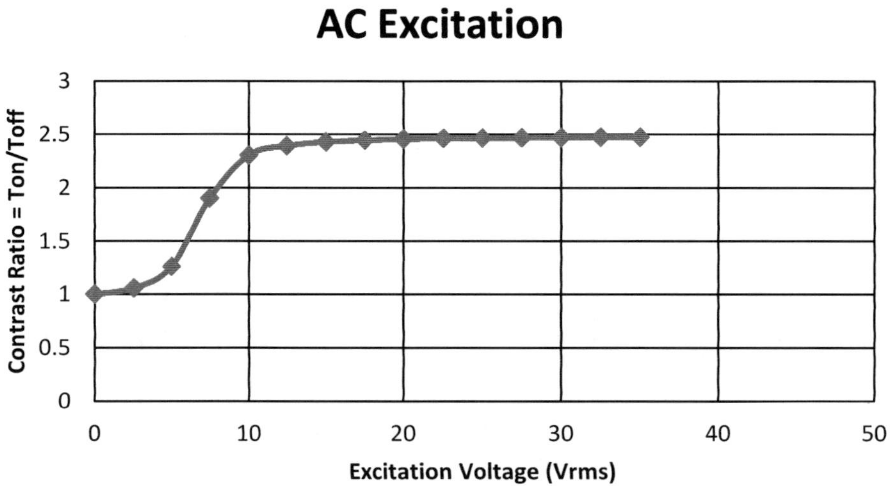

Figure 7. Contrast ratio of the PDLC under AC 50 Hz excitation.

Fig. 8 and 9 display comparison between optical properties of the PDLC under DC and AC excitation, where we observed a similar T-V curve and contrast ratio, except small differences at the transition points and slightly higher maximum contrast ratio by DC excitation compared to AC excitation.

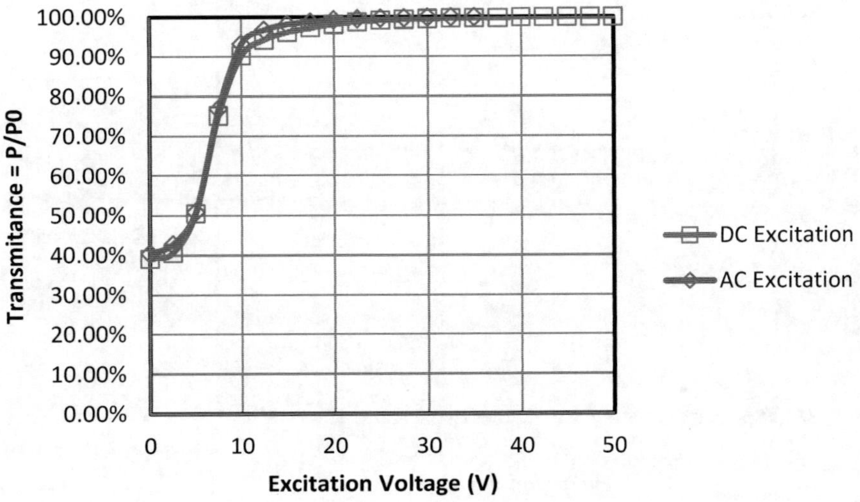

Figure 8. Comparison between transmittance of the PDLC under DC and AC 50 Hz excitations.

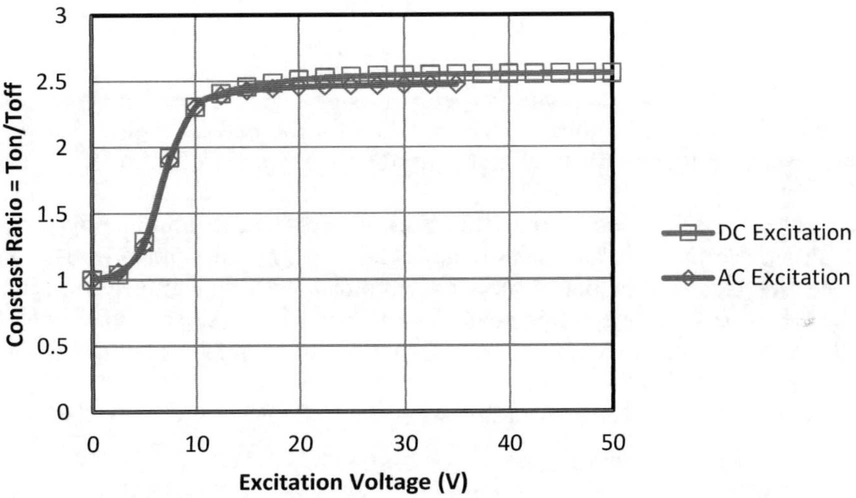

Figure 9. Comparison between contrast ratio of the PDLC under DC and AC 50 Hz excitations.

3.3. Electrical measurement under DC and AC excitation

The basic electrical characteristics of the PDLC was also measured, which results in no steady state current flow across the PDLC under DC excitation, as the device is capacitive. However, when excited by AC source, the PDLC consumes steady state reactive power. Although this power will not be counted by the energy meter, but it will deteriorate the power factor. Fig. 10 shows that the I-V curve of the PDLC under DC and AC excitations are very much different. The slope of the curve under AC excitation shows that the device has admittance of 0.0901 mS or impedance of 11.1 kOhm leading to capacitance of 0.2868 μF and capacitance per area of 18.5 μF/m^2. As many applications of PDLC in the field use AC excitation as it is easily accessible and transformed at typical installation spots, these results show that DC excitation offers advantage in term of power factor. The power factor change for AC excitation will also scale with the area size of the PDLC.

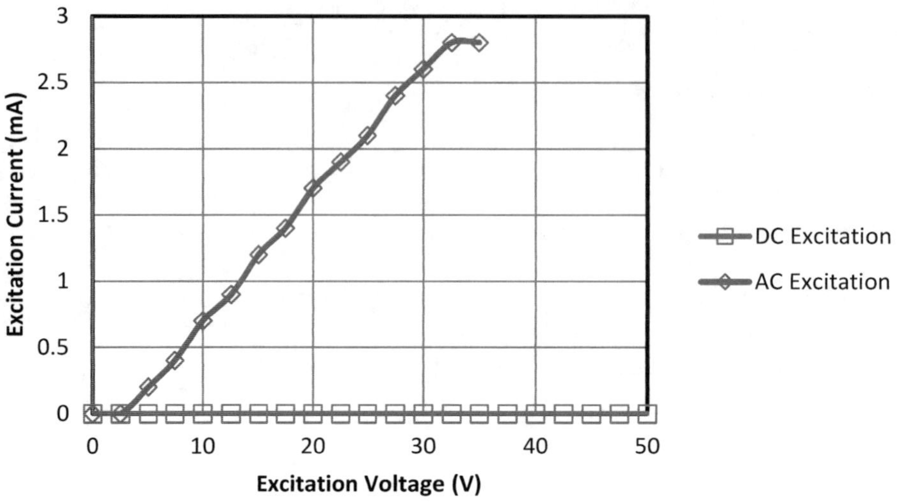

Figure 10. Comparison between electrical characteristics of the PDLC under DC and AC 50Hz excitations.

4. Design of Interface to Arduino

4.1. DC/DC boost circuit with Arduino interface

In order to demonstrate the applications of DC excitation for the PDLC for smarthome, we designed a DC/DC boost converter that takes 5V (commonly used voltage in microcontroller system) as input dan produces DC output of programable voltage from the low voltage to exceeding V_{90} of the PDLC.

Fig. 11 shows the schematic of the circuit, where the NE555 multiple-purpose timer chip is configured as an astable multivibrator with tunable duty cycle (and simultaneously also frequency) through the potentiometer. The output of the multivibrator then through a selection jumper P_1 drives transistor Q_1 which behaves as electronic switch. The transistor together with L_1, D_1, D_2, C_3 and C_4 will form a DC/DC boost converter. The circuit is intended to be able to give DC 6 –30 V to fed into the PDLC electrodes for testing purposes. Fig. 12 shows the output pulses of the timer chip, with Fig. 12.a associated to 7 V DC output and Fig. 12.b associated to 27.9 V which is well above the V_{90} of the PDLC.

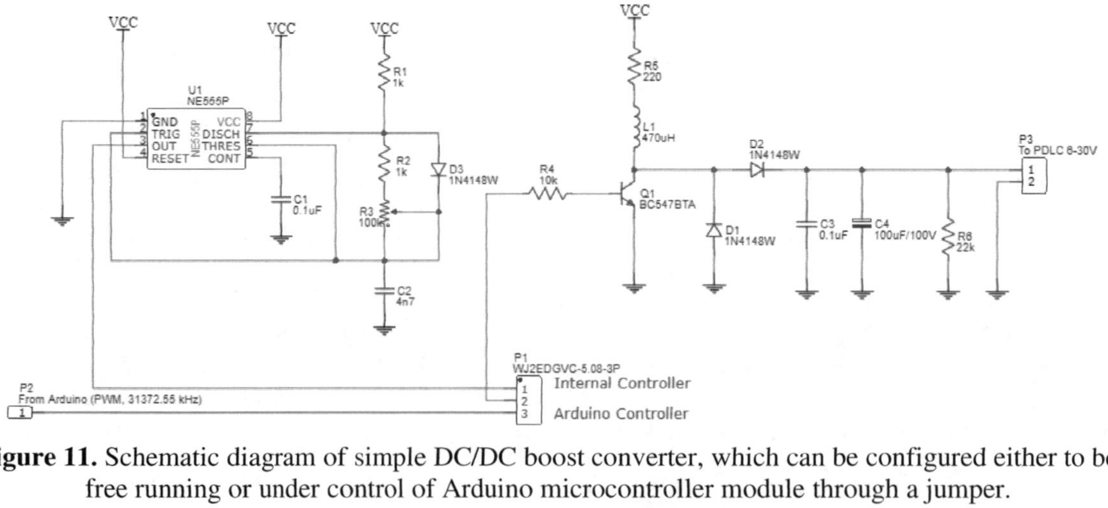

Figure 11. Schematic diagram of simple DC/DC boost converter, which can be configured either to be free running or under control of Arduino microcontroller module through a jumper.

Journal of Physics: Conference Series **2274** (2022) 012010 doi:10.1088/1742-6596/2274/1/012010

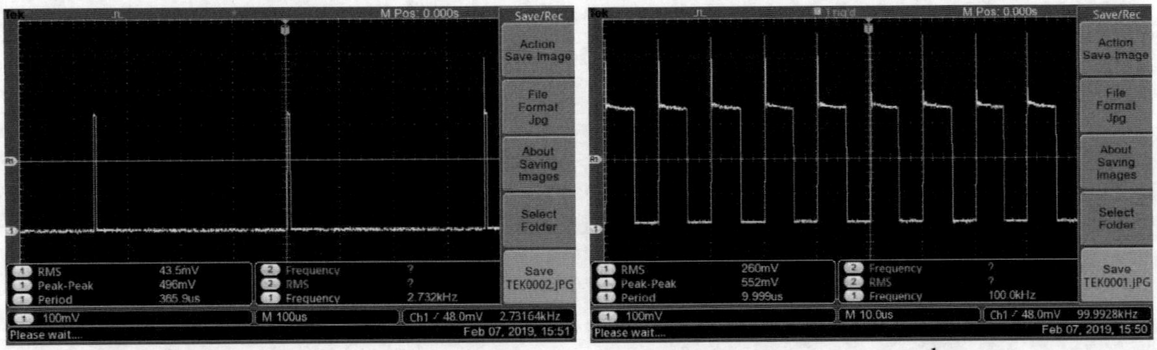

a. b.

Figure 12. The output pulses of NE555 when: a. the potentiometer gives lowest resistance, and b. the potentiometer gives highest resistance.

4.2. Measurement under Arduino control

In order to make the PDLC to be part of a smarthome, it should be controllable by a microcontroller system. For this, we configured jumper P_1, so that Q_1 get controlled from a pulse width modulation (PWM) output port of an Arduino [8] microcontroller module as shown in Fig. 13. In this way, the DC/DC boost converter output is determined by the PWM pulse output of the Arduino which is set to a frequency of 31372.55 Hz through its prescaler. By changing the duty cycle of the PWM output of the Arduino, the PDLC excitation voltage changes accordingly as shown in Fig. 14, which results in changes of transmittance of the PDLC as shown in Fig.15. We note that with duty cycle of 0%, the output of the DC/DC converter still gives 4.4 V which can be understood from schematic diagram in Fig. 11, that when Q_1 is off, there is still voltage supplied from V_{cc} through R_5, L_1, dan D_2 to the PDLC. With this configuration, by changing the duty cycle of the PWM from 0 to 39% will results into contrast ratio of 2.05 for the PDLC. This 39% of duty cycle already produces voltage well beyond the driving voltage of the PDLC. We should note, that duty cycle of 100% should be avoided to prevent damaging the transistor as can understood from schematic diagram in Fig. 11.

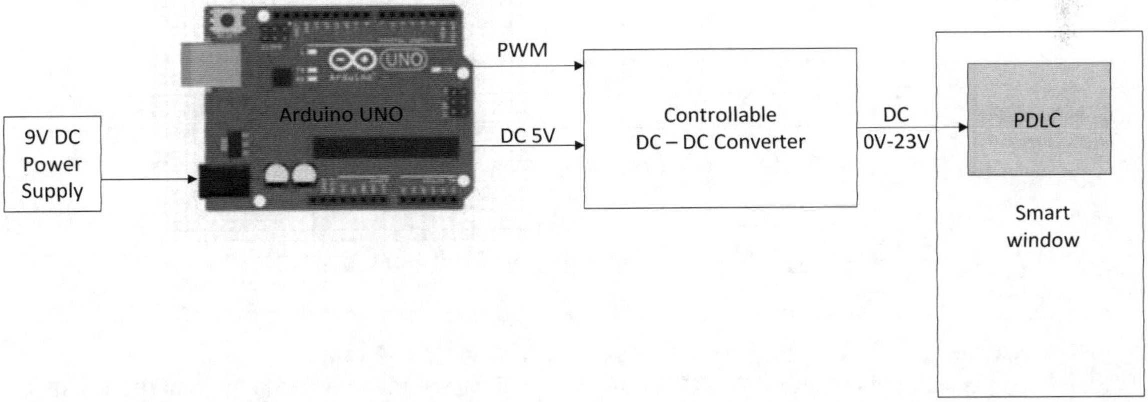

Figure 13. The configuration for controlling the DC excitation voltage of the PDLC through Arduino.

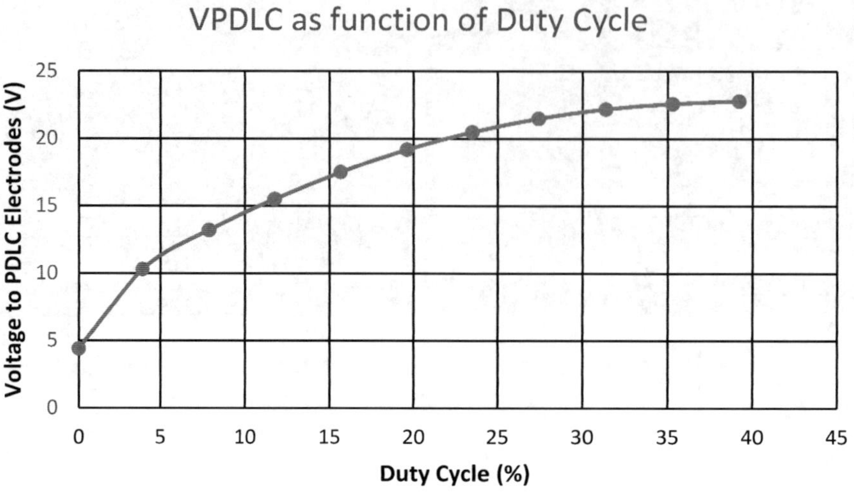

Figure 14. The measured DC excitation voltage to the PDLC as controlled by the PWM duty cycle of the Arduino.

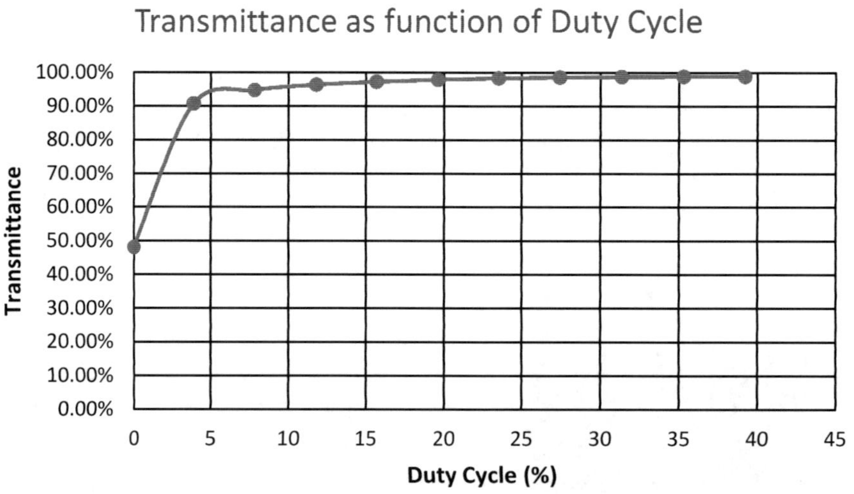

Figure 15. The measured transmittance of the PDLC as controlled by the PWM duty cycle of the Arduino.

5. Demonstration of PDLC for Smart Window with Voice Command

To demonstrate the application of PDLC for smart window, we made a miniature smarthome model shown in Fig. 16, using scheme shown in Fig. 17, where the PDLC as smart window can be controlled to be made transparent and white translucent by human voice command recognized by Google Assistant [9]. In this scheme, the voice command from human was recognized by the Google Assistant with client code embedded in an Android smartphone, and according to certain programmed phrases, if recognized, the Google Assistant server will pass it to the IFTTT [10] logical server which will set the value of certain command variables in the Blynk [11] IoT server. The Arduino as IoT node through ESP8266 WiFi transceiver module will check the IoT server periodically through home's network connected to the internet, and based on the value of the command variables, the Arduino will generate PWM signal with corresponding duty cycle to make the PDLC transparent or translucent. The details of this miniature smarthome model were already discussed in Adhinugroho and Uranus [4]. Figure 18 shows the window while being commanded to be in translucent and transparent state.

Figure 16. The miniature smarthome model in which the PDLC is implemented for a smart window.

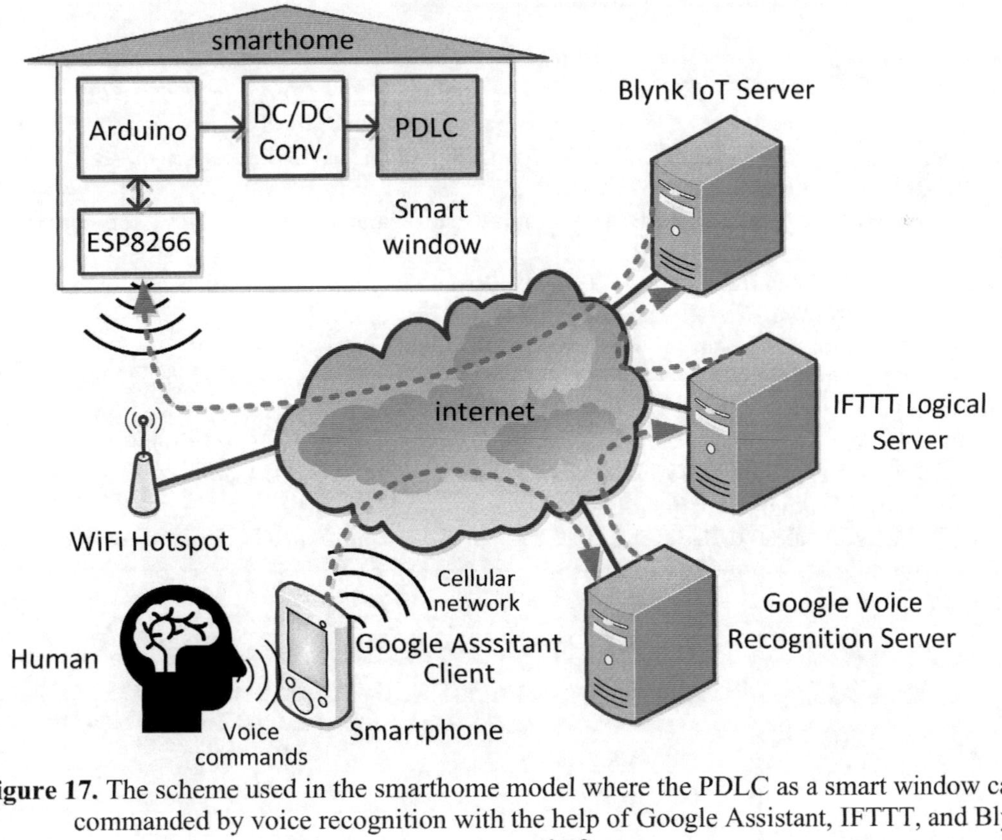

Figure 17. The scheme used in the smarthome model where the PDLC as a smart window can be commanded by voice recognition with the help of Google Assistant, IFTTT, and Blynk IoT platform.

a. b.

Figure 18. The PDLC as a smart window in: a. translucent state, and b. transparent state.

6. Conclusions

A commercial PDLC sample was characterized for its basic optical and electrical properties. The results were used to develop a DC excitation circuit for implementation of the PDLC as a smart window in a miniature smarthome model. The results show that excitation of PDLC using DC is feasible while offering advantage of lower steady state power consumption and higher possible contrast ratio compared to commonly used AC excitation.

Acknowledgments

This work was carried out under Project Penelitian Internal Mandiri No. P-022-M/FaST/X/2021 supported by LPPM Universitas Pelita Harapan.

References

[1] Nugroho T A, Suakanto S, Sutarto H Y, and Joelianto E 2019 *Internetworking Indonesia J.* **11** 29
[2] Alaa M, Zaidan A A, Zaidan B B, Talal M, and Kiah M L M. 2017 *J. Network & Comp. Apps.* **97** 48
[3] Rešetič A, Milavec J, Zupančič B, Domenici V, and Zalar B. 2016 *Nature Communications* **7** 13140
[4] Adhinugroho N R and Uranus H P 2019 *Proc. Seminar Nasional Sains, Rekayasa, dan Teknologi (SNSRT) 2019*
[5] Lee Y H, Cheruku R, Thogiti S, and Kim J H 2016 *Sci. Adv. Mat.* **8** 1
[6] Malik M K, Bhati P G, and Deshmukh R R 2012 *Indian J. Science & Technol.* **5** 3440
[7] Amanah D and Harahap D A 2020 *European J. Business and Management Research* **5** 1
[8] Badamasi Y A 2014 *11th International Conf. Electronics, Computer and Computation*
[9] Gupta M P 2018 *International Research J. Eng. and Technol.* **5** 2074
[10] Ovadia S 2014 *Behavioral & Social Sciences Librarian* **33** 208
[11] Media's E, Syufrijal, and Rif'an M 2018 *3rd UNJ International Conference on Technical and Vocational Education and Training*

The effect of immersion time on singlemode-tapered multimode-singlemode (STMS) fabrication using a chemical etching method

A. D. Lesa [1], I. Puspita[1*], A. M. Hatta[1], F. Kurniawan[2] and S. Koentjoro[1]

[1] Department of Engineering Physics, Institut Teknologi Sepuluh Nopember, Surabaya, Indonesia
[2] Department of Chemistry, Institut Teknologi Sepuluh Nopember, Surabaya, Indonesia

*Corresponding author: ika.tf10@gmail.com

Abstract. Singlemode-Multimode-Singlemode (SMS) is one of the fiber optic sensor structures that is widely applied for many sensing applications. High sensitivity is the preferred characteristic of SMS fiber structure. In this paper, a tapered structure is introduced to the SMS fiber structure to increase its performance. The tapered structure is made by the etching method by immersing the SMS multimode fiber in hydrofluoric acid (HF) solution. The immersion time was varied 5, 10, 15 and 20 minutes. The diameter of the multimode fiber decreases with the longer immersion time, while the length does not change. The STMS transmission spectrum was measured to investigate its performance. The chemical etching method using HF solution for tapering the SMS fibers has great potential as a simple method to achieve a smaller radius without lengthening multimode fibers.

1. Introduction

Singlemode-Multimode-Singlemode (SMS) is one of the modified structures of optical fiber sensors and has been widely applied for many sensing applications, physical and chemical/bio sensor. It has several advantages including non-destructive, high sensitivity, easy fabrication and installation, immunity to electromagnetic interference, compact, and low cost of fabrication [1]. SMS is fabricated by axially splicing a multimode fiber (MMF) between two identical single-mode fibers (SMFs) [2]. As a sensor, high sensitivity is a preferable characteristic. The fabrication results [3] show that the SMS structure has great potential as a sensor by producing a greater sensitivity than the sensitivity of the SMF and MMF structures.

The sensitivity of optical fiber can be increased by making the fiber into a taper structure [4]. Taper optical fibers provide some benefits such as generating large evanescent fields, having strong mode confinement capabilities and small diameters. Taper optical fibers show great potential for applications in optical communication and sensing systems [5]. The performance of the taper optical fiber depends on the taper length, waist diameter and interaction length.

The tapered structure on the SMS fiber structure can be fabricated by mechanical method or chemical method [6], [7]. Mechanical methods are carried out by heat pulling the fiber while the chemical method is by immersion MMF region into chemicals such as hydrofluoric acid (HF) solution which is corrosive and can reduce the diameter of the MMF core. Fabrication of taper using a laser-heating has been reported to produce excellent tapers. However, it is expensive. The chemical etching method can be used to fabricate a taper structure since it is easily fabricated and simple method [4]. Therefore, in this paper the chemical etching method was chosen in the fabrication of SMS tapers where the effect of immersion time on Singlemode-Tapered Multimode-Singlemode (STMS) was then investigated.

2. Basic Theory

It is well-known that the STMS structure is an extension of the SMS structure, therefore the wave modal propagation within the MMF becomes very useful for the design of STMS. The taper optical fiber has three main parts, namely the untapered region, the transition region, and the taper waist region as shown in Figure 1. In the untapered region of MMF fiber optics, the thin core is surrounded by a cladding with a lower refractive index. Light is guided in the core due to a total internal reflection. The light properties in the fiber core are determined by the number of modes that are directly related to the V number, which is called as the normalized frequency parameter given by:

$$V = \frac{2\pi}{\lambda} r \sqrt{{n_{co}}^2 - {n_{cl}}^2} \tag{1}$$

The guided modes exist in the transition region. However, in the transition region, the core-cladding arrangement is almost mixed as the diameter decreases. In such situation it can be regarded as a core that is surrounded by air, which results in a large V number due to a larger difference in refractive index between the core and air. In the transition region, there are two conditions, namely adiabatic and nonadiabatic. In the adiabatic condition, the efficiency can reach as high as 99.5% [8]. On the other hand, in the non-adiabatic condition, high-order modes can be excited into an evanescent field. In the taper waist area, the V number of the optical fiber changes according to the change in the diameter of the optical fiber.

Some constructive interference phenomena may occur in the taper MMF section due to the focusing effect of the left taper transition region. In the taper section of the MMF, the number of modes and the intensity of the interference of modes in the right side of the taper is much smaller and lower than those of the conventional SMS structure, due to the high-order mode filtering effect caused by the taper structure. At the same time, the silica cladding layer still exists outside the MMF taper core between the MMF core and the substance to be absorbed; the confinement effect which is provided by this thin cladding layer can extend the dynamic range of the STMS refractometer [6].

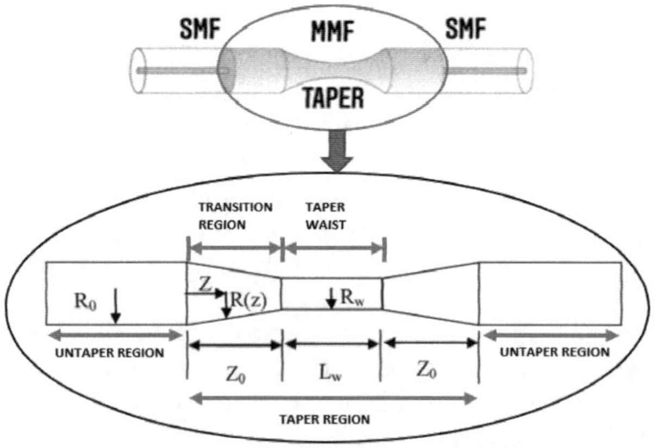

Figure 1. Structure of STMS fiber

In the previous work by Wang *et al.*, it was explained that the guided mode propagation was the starting point of reference. [9] When the light propagates from the SMF structure into the MMF, the fundamental mode $E(r, 0)$ are decomposed into a number of M eigen modes $\phi_m(r)$ so that

$$E(r,0) = \sum_{m=1}^{M} a_m \phi_m(r) \tag{2}$$

where m and a_m represent the excited mode number and the excitation coefficient, respectively. The two parameters can be calculated from:

$$a_m = \frac{\int_0^\infty E(r,0)\phi_m(r)rdr}{\int_0^\infty \phi_m(r)\phi_m(r)rdr} \tag{3}$$

and

$$M \approx \frac{V}{\pi} = \frac{2a}{\lambda}\sqrt{{n_{co}}^2 - {n_{cl}}^2} \tag{4}$$

where a is the radius of the MMF core, n_{co} and n_{cl} are the refractive index of the MMF core and cladding. The optical amplitude distribution of the transmitted light over a propagation distance z in the MMF section can be expressed as

$$E(r,z) = \sum_{m=1}^{M} a_m \phi_m(r) \exp{(i\beta_m z)} \tag{5}$$

where β represents of mth mode's propagation constant. Based on this equation, the optical amplitude distribution pattern at propagation distance z within the SMS Fiber structure can be calculated [9]. When light is launched from the MMF section into the output SMF, the eigenmodes within the MMF are recoupled into the SMF fundamental mode and the relationship between the coupling loss and the MMF length can be calculated from:

$$L_s(z) = 10log_{10}\left(\left|\sum_{m=1}^{M} {a_m}^2 \exp(i\beta_m z)\right|^2\right) \tag{6}$$

The concept of mode guidance described above is very important in the implementation to design the STMS structure shown in Figure 1. The STMS consists of multimode fiber (MMF) that is located between two identical single-mode fibers (SMFs). The left section of the SMF is used as the input arm, where the SMF input is the fundamental mode, and the right SMF is used as the light output arm. Here, the design is focused on the MMF section, as it is related to efforts to increase the sensitivity of the STMS. The geometric parameters of MMF, namely the taper waist $R(z)$, and the transition length z_0 are connected by the following equation [6]:

$$R(Z) = R_0 \exp\left(\frac{-Z}{L_0}\right) \tag{7}$$

and

$$Z_0 = L_\omega \ln\left(\frac{R_0}{R_\omega}\right) \tag{8}$$

where R_0 is the initial radius of the uniform MMF, R_ω and L_ω are the radius and length of the taper waist area.When the parameters of the tapered multimode fiber are substituted into the above equations, we obtain the optical amplitude distributions of the propagation fields within the STMS fiber structure and also the relationship between the coupling loss at the output SMF and the tapered MMF length.

3. Methods

The fabrication of Singlemode-Taper Multimode-Singlemode (STMS) optical fiber is initiated by the preparation of the two single-mode optical fibers (SMF) and afterward the preparation of the one multimode optical fiber (MMF). The detailed specifications of the fiber were given in Table 1. The cladding was peeled off using a striper and cut along 1 cm using a Fujikura 6s cleaver. STMS optical fiber is made by connecting two single-mode optical fibers (SMFs) and multimode optical fibers (MMF) using Sumitomo Z1C fiber splicers. STMS fabrication is done by peeling the fiber cladding and reducing the diameter of the core of MMF from the input side to the waist and then increasing again to the output side, as illustrated in Figure 1. The taper structure is made by laying the SMS structure in an acrylic mold. Inside the acrylic mold, there is a concave container that contains HF solution as the immersion place of the MMF section as shown in Figure 2. The immersion time was varied 5 minutes, 10 minutes, 15 minutes, and 20 minutes.

The spectrum of the fabricated STMS was measured using a spectrometer by OTO Photonics in the range of 900 – 1600 nm. An optical laser source was utilized as the light source at the operating wavelength of 1550 nm.

Table 1. Optical fiber specifications

Parameters	Singlemode pigtail	Multimode pigtail	Multimode coreless
Core diameter	8.3 μm	105 μm	-
Cladding diameter	125 μm	125 μm	125 μm
Core refractive index	1.45	1.44	-
Cladding refractive index	1.44	1.42	1.45

Figure 2. Immersion process using HF solutions in acrylic mold

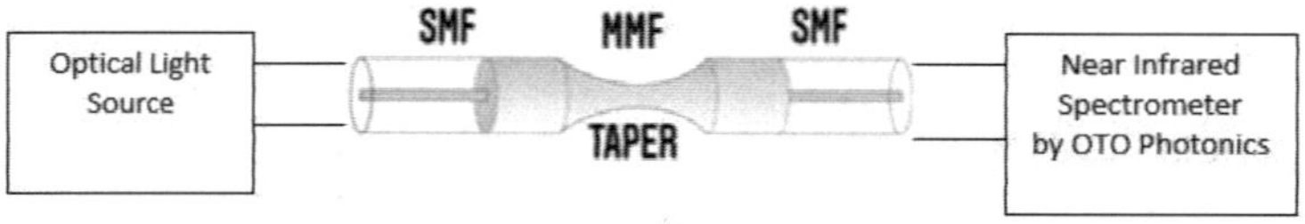

Figure 3. Schematic of spectrum-based STMS optical fiber sensor characterization

4. Results and Discussion

The following is the result of immersion in HF solution based on the immersion time. In Figure 4, there are 4 results of SMS fiber optic fabrication, the first is the uniform structure where in this structure no treatment is carried out on the MMF. Then the STMS 5, STMS 10, STMS15 where each part of the MMF section of the SMS structure is immersed in HF solution with immersion time for 5 minutes, 10 minutes and 15 minutes, respectively.

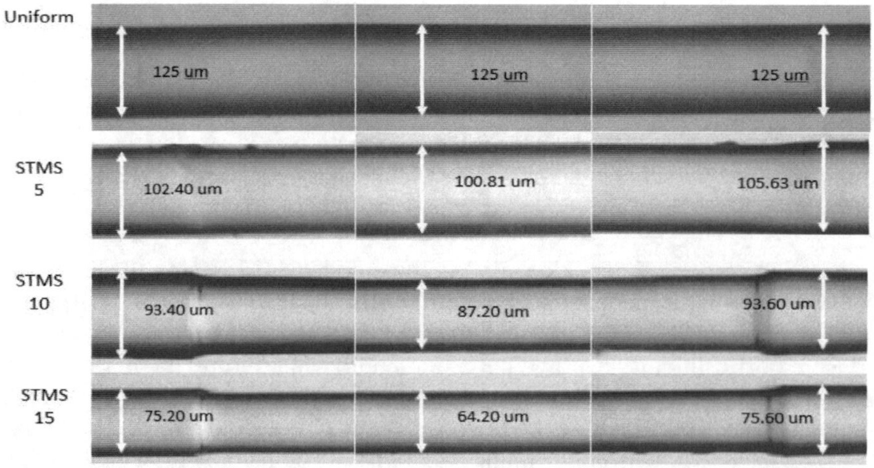

Figure 4. Microscopic image of STMS fiber with the varied time of immersion

Figure 4 shows the microscopic image of the STMS fiber structure with the variation of immersion time. One can see that the diameter of the multimode fiber reduced as the longer immersion time while its length does not change. The optimum diameter can be obtained is 62.4 μm by immersing in HF solution for 15 minutes. The smallest diameter can be obtained after 20 minutes immersed in HF solution. However, the optical fiber reached its strength limit. Hence, it breaks after immersion time. Thus, it cannot be immersed for more than 15 minutes.

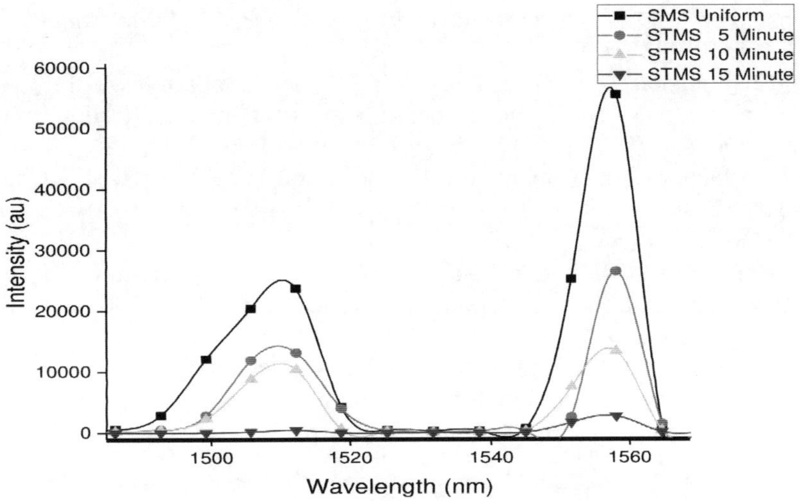

Figure 5. Transmission spectrum of STMS fiber structure

The STMS transmission spectrum was measured using a near-infrared spectrometer by OTO-Photonics. The transmission spectrum of STMS at the wavelength range of 1495 – 1565 nm is depicted in Figure 5. One can see, that for pre-etching structure, an interference phenomenon occurred in the SMS fiber structure, which is indicated by periodic peaks measured in its spectrum. By introducing a taper structure on the SMS fiber structure, the intensity of each wavelength peak decreased with the longer immersion time of etching. The smaller the diameter of the taper waist on STMS, the filtering effect did occur. This result has a great agreement with the previous work [6].

5. Conclusions

The STMS fiber optic fabrication has been demonstrated. From the experimental results, it can be concluded that the longer immersion time gives the smaller diameter and the lower output intensity. This is related to the filtering effect in the MMF region. It can be concluded also that the chemical etching method using HF solution for tapering the SMS fiber structure has great potential as a simple method to achieve a smaller radius without lengthening the multimode fiber and can be utilized for many sensing applications.

Acknowledgments

This work was supported by the Ministry of Education, Culture, Research and Technology under Grant No. 918/PKS/ITS/2021.

References

[1] Wu Q. *et al.*, "Singlemode Multimode-Singlemode Fiber Structures for Sensing Applications-A Review," *IEEE Sens. J*, vol. 21, no. 11, pp. 12734–12751, 2021, doi: 10.1109/JSEN.2020.3039912.

[2] Puspita I., Rahmah F., Hatta A. M., and Koentjoro S., "Load effect on an SMS fiber structure embedded in a high-density polyethylene," *Int. Semin. Photonics, Opt. Its Appl. (ISPhOA*

2014), vol. 9444, no. 2, p. 94440W, 2015, doi: 10.1117/12.2083343.

[3] Puspita I. *et al.*, "Detection of Lard Adulteration in Olive Oil by Using Silica Optical Fiber," *ISSIMM 2018 - 3rd Int. Semin. Sensors, Instrumentation, Meas. Metrol. Proceeding*, pp. 119–122, 2018, doi: 10.1109/ISSIMM.2018.8727640.

[4] Haddock H. S., Shankar P. M., and Mutharasan R., "Fabrication of biconical tapered optical fibers using hydrofluoric acid," *Mater. Sci. Eng. B Solid-State Mater. Adv. Technol.*, vol. 97, no. 1, pp. 87–93, 2003, doi: 10.1016/S0921-5107(02)00434-8.

[5] Wang P., Zhao H., Wang X., Farrell G., and Brambilla G., "A review of multimode interference in tapered optical fibers and related applications," *Sensors (Switzerland)*, vol. 18, no. 3, 2018, doi: 10.3390/s18030858.

[6] Wang P., Brambilla G., Ding M., Semenova Y., Wu Q., and Farrell G., "Investigation of single-mode – multimode – single-mode fiber structures and their application for refractive index sensing," *America (NY).*, vol. 28, no. 5, pp. 1180–1186, 2011.

[7] Kbashi H. J., "Fabrication of Submicron-Diameter and Taper Fibers Using Chemical Etching," *J. Mater. Sci. Technol.*, vol. 28, no. 4, pp. 308–312, 2012, doi: 10.1016/S1005-0302(12)60059-0.

[8] Tian Y., Wang W., Wu N., Zou X., and Wang X., "Tapered optical fiber sensor for label-free detection of biomolecules," *Sensors*, vol. 11, no. 4, pp. 3780–3790, 2011, doi: 10.3390/s110403780.

[9] Puspita I., Banurea D. P. M., Irawati N., Hatta A. M., Sekartedjo, and Kurniawan F., "Taper parameters effect on tapered POF for lard adulteration in olive oil detection," *Optoelectron. Adv. Mater. Rapid Commun.*, vol. 14, no. 5–6, pp. 250–255, 2020.

Journal of Physics: Conference Series **2274** (2022) 012012 doi:10.1088/1742-6596/2274/1/012012

Refractive index sensing performance of a bimetallic core-shell nanotube

Azrul Azwar*

Department of Physics, Faculty of Mathematics and Natural Sciences, Tanjungpura University, Indonesia

*a.azwar@physics.untan.ac.id

Abstract. The sensing performance of a bimetallic core-shell nanotube is studied theoretically based on the theory of TE-polarized light scattering by a multilayer infinite cylinder. The nanotube consists of a metallic core and a metallic shell with the analyte is inserted between the core and the shell. It is found that the absorption efficiency spectra of a bimetallic core-shell nanotube exhibit three resonance peaks correspond to the dipole, quadrupole and sextupole mode. As the refractive index of the analyte increased from n_2 = 1.33 to n_2 = 1.45, these resonance peaks are linearly redshifted. By calculating the figure of merit (FOM), it is found that the high sensing performance with FOM > 100 could be achieved. The dipole mode exhibits a higher sensing performance as compared to the other modes. Finally, the sensing performance is very sensitive to variations of the geometrical parameter and the background medium of the nanotube.

1. Introduction

Localized surface plasmon resonances (LSPRs) are optical phenomena that occur in metallic nanostructures in which collective charge oscillations are induced at metal-dielectric interfaces. The unique properties of LSPRs such as enhanced absorption or scattering cross section make them suitable in a wide ranging applications. All of these unique properties are very sensitive to the metal type, geometrical structure, and the refractive index of the surrounding medium [1]. In recent years, many studies have been conducted to explore the effect of the system's parameters variations to the optical response of plasmonic nanostructure for specific application. One of the most important applications of LSPRs is refractive index sensing which exploit the changing of the LSPRs properties due to the refractive index variation of the medium to be sensed [2].

It is known that, the refractive index sensing ability of the metallic nanostructure could be improved by using the core–shell structure, due to the coupling of the plasmon modes. For example, Tam et al. have reported that the sensitivity could be enhanced by fabricating larger gold nanoshell with thinner shell thickness [3]. Further, Zhu investigated the core refractive index sensitivity of LSPRs by using the quasi-static calculation in gold nanotube and found that the core refractive index sensitivity increased as the decreased of wall thickness or the background refractive index [4]. Other structural configuration that has potential for index sensing application is the nanowire system. It has been shown that this nanowire system possesses the asymmetric Fano resonance response [5, 6] that can be tuned sensitively by its geometrical parameters. Among the various nano cylindrical structure, the tubular nanostructure is now become interesting, because it has open ended surfaces which has a potentially as a cavity for index sensing [7] as well as a microreactor. Using the microfluidic and nanofluidic technique, this cavity or microreactor could be filled by the chemical substance to produce a chemical reaction in nanoscale. This nanoscale chemical reaction could be investigated by the changing of the dielectric properties of material inside the tubular structure. Experimentally, this type of nanostructure was successfully fabricated by using a self-assembly technique and found that this vertically array gold nanotube could be used as a high performance refractive index sensor [8]. In order to improve the sensing performance of the nanotube, we propose in this study a modified tubular nanostructure by adding a silver core inside the silver nanotube with the analyte is located in the space between core and shell. We found that this structure exhibits geometrical tunability and could boost the figure of merit 10 times higher than the nanotube without core [9].

2. Theoretical Formulation and Sensing Principles

In this study, the nanotube is modelled as a concentric multi-layer infinite cylinder consisting of a cylindrical metallic core with radius R_1 and metallic shell with the inner and outer radii denote by R_2 and R_3, respectively. The analyte is placed between the core and the shell as shown in Fig. 1. The refractive indices of the analyte ($n_2 = \sqrt{\varepsilon_2}$) and background medium ($n_4 = \sqrt{\varepsilon_4}$) are assume to be real and non-dispersive, while refractive indices of the metallic core ($n_1 = \sqrt{\varepsilon_1}$) and shell ($n_3 = \sqrt{\varepsilon_3}$) are complex and dispersive which taken from the realistic experimental data of silver [10]. The schematic diagram of the TE-polarized light scattering, and the configuration of the analyte inside the nanotube is given in Figure 1

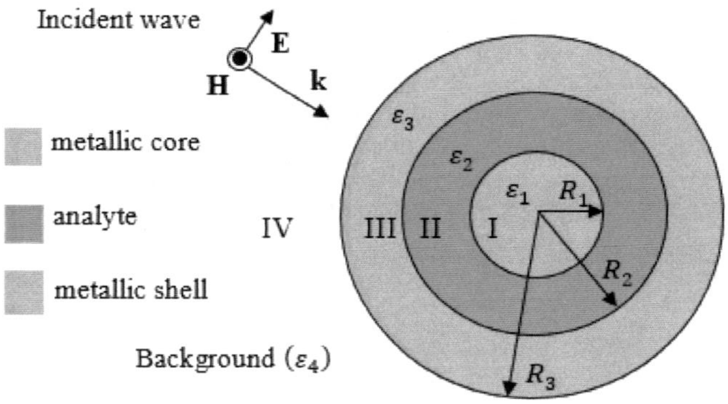

Fig. 1. Schematic description of the analyte inside a bimetallic core–shell nanotube and the TE-wave scattering configuration

Using the standard formulation of TE polarized electromagnetic scattering by a multilayer infinite cylinder for the plane wave of normalized amplitude propagates in a direction perpendicular to the nanotube axis [11], the general solution for the magnetic field in each region are given below

Region I ($r < R_1$) $$H_z^I = -\sum_{n=-\infty}^{\infty} (-i)^n n_1 [f_n J_n(k_1 r)] e^{in\varphi} \quad (1)$$

Region II ($R_1 < r < R_2$) $$H_z^{II} = -\sum_{n=-\infty}^{\infty} (-i)^n n_2 \left[d_n J_n(k_2 r) + e_n H_n^{(1)}(k_2 r)\right] e^{in\varphi} \quad (2)$$

Region III ($R_2 < r < R_3$) $$H_z^{III} = -\sum_{n=-\infty}^{\infty} (-i)^n n_3 \left[b_n J_n(k_3 r) + c_n H_n^{(1)}(k_3 r)\right] e^{in\varphi} \quad (3)$$

Region IV ($r > R_3$) $$H_z^{IV} = -\sum_{n=-\infty}^{\infty} (-i)^n n_4 \left[J_n(k_4 r) - a_n H_n^{(1)}(k_4 r)\right] e^{in\varphi} \quad (4)$$

where $k_i = k_0\sqrt{\varepsilon_i}$ with k_0 is wave number in vacuum, while J_n and $H_n^{(1)}$ are the Bessel and the first kind Hankel functions, respectively. By inserting the Eq. (1) – (4) to the Maxwell equation, one can find the respective electric field. Applying the boundary conditions of tangential components of the electric and magnetic fields at each interface of the cylinder gives the following relations

$$f_n J'_n(k_1R_1) - d_n J'_n(k_2R_1) = e_n H'^{(1)}_n(k_2R_1) \quad (5)$$

$$f_n n_1 J_n(k_1R_1) - d_n n_2 J_n(k_2R_1) = e_n n_2 H_n^{(1)}(k_2R_1) \quad (6)$$

$$b_n J'_n(k_3R_2) + c_n H'^{(1)}_n(k_3R_2) - d_n J'_n(k_2R_2) = e_n H'^{(1)}_n(k_2R_2) \quad (7)$$

$$b_n n_3 J_n(k_3R_2) + c_n n_3 H_n^{(1)}(k_3R_2) - d_n n_2 J_n(k_2R_2) = e_n n_2 H_n^{(1)}(k_2R_2) \quad (8)$$

$$a_n H'^{(1)}_n(k_4R_3) + b_n J'_n(k_3R_3) + c_n H'^{(1)}_n(k_3R_3) = J'_n(k_4R_3) \quad (9)$$

$$a_n n_4 H_n^{(1)}(k_4R_3) + b_n n_3 J_n(k_3R_3) + c_n n_3 H_n^{(1)}(k_3R_3) = n_4 J_n(k_4R_3) \quad (10)$$

Solving the linear system of Eq. (5) – (10), one can find the multipole expansion coefficient a_n, b_n, c_n, d_n, e_n and f_n.

The excitation of LSPRs can be observed and identified from the absorption of the incoming electromagnetic wave energy which is presented by the absorption efficiency Q_{abs} that is expressed in the terms of multipole expansion coefficient given by

$$Q_{abs} = \frac{2}{(k_4R_3)} \sum_{n=-\infty}^{\infty} \{Re\,(a_n) - |a_n|^2\} \quad (11)$$

The rate of change of the LSPR peaks (λ_r) due to the variation of analyte refractive index gives the definition of sensitivity (S),

$$S = \frac{d\lambda_r}{dn_2} \quad (12)$$

while the related sensing performance is characterized by the Figure of Merit (FOM) defined by

$$\text{FOM} = \frac{S}{\Delta\lambda} = \frac{1}{\Delta\lambda}\left(\frac{d\lambda_r}{dn_2}\right) \quad (13)$$

where $\Delta\lambda$ is the width of the related spectral profile defined as the FWHM (Full Width at Half Maximum).

3. Results and Discussion

3.1. Absorption Efficiency

The LSPRs phenomena could be investigated in the resonance feature of the absorption efficiency (Q_{abs}) spectra. Using Eq. (11), one can calculate the spectral variation of Q_{abs} for core-shell nanotube with R_1 = 50 nm, R_2 = 75 nm and R_3 = 125 nm and the result is given in Fig. 2 (a) and the permittivities of core-shell nanotube structure are given in the caption. It is seen that in the region of visible to near-infrared, the profile of Q_{abs} exhibits three distinct peaks as shown in Fig. 2(a). The peaks of the longest wavelength (λ = 896.0 nm) could be identified as a dipole mode resonance (D) as indicated in the electric field intensity $|\mathbf{E}|^2$ distribution near the nanotube shown in Fig. 2(b). Since the size of the nanotube is large enough, the higher resonance modes with shorter wavelength could be exited. As demonstrated in Fig. 2(c), the spatial distribution of $|\mathbf{E}|^2$ at λ = 509.0 nm indicates a quadrupole mode resonance (Q), while resonance of the shortest wavelength (λ = 411.0 nm) exhibits a sextupole mode resonance (S) feature as shown in Fig. 2(d). Further, one observes that at the resonances condition, the field distribution is confined in the analyte region. The field confinement indicates the effective coupling between surface plasmons of core and inner surface of the shell. This coupling of surface plasmons was due to the hybridization of the resonance, similar to plasmon hybridization of a nano cylindrical shell [11], where it was shown that the plasmon resonance depends sensitively to the analyte refractive index. The existence of these three resonance modes with highly localized field intensity distribution could be exploited for the multiband sensing application with ultra-high figure of merit as will be discussed in the following sections.

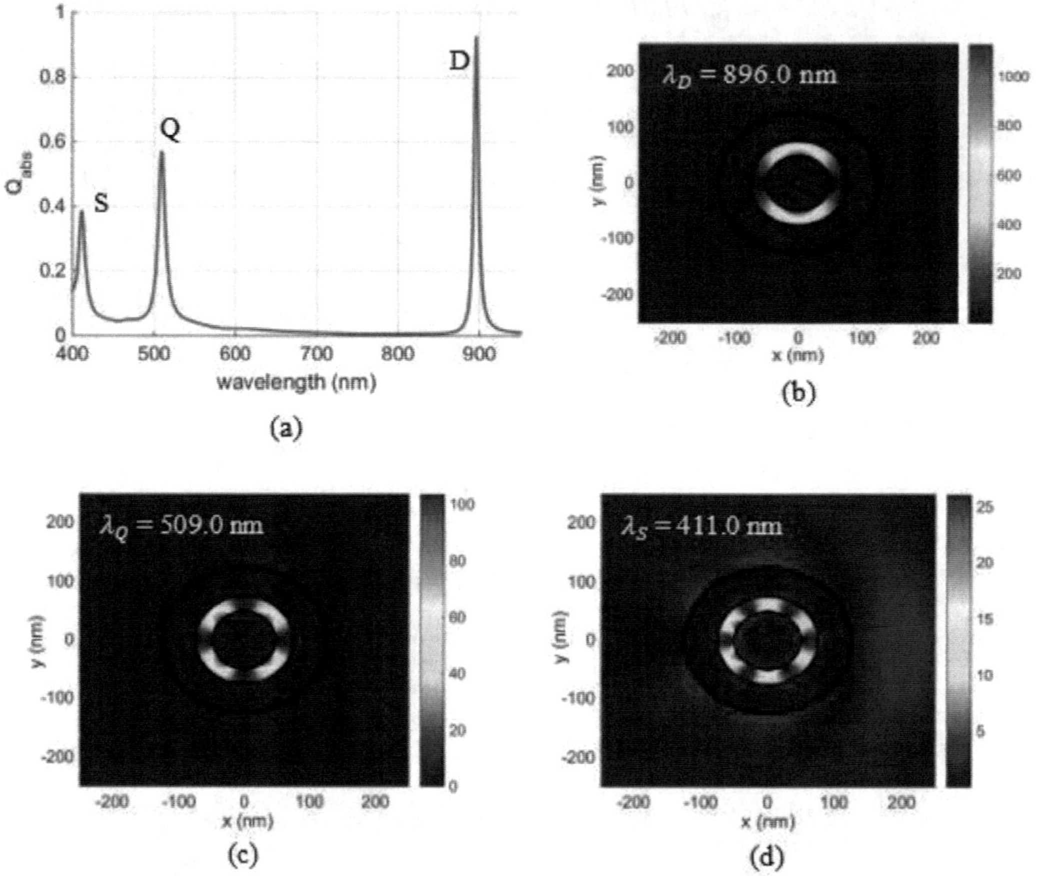

Fig. 2. (a) The wavelength dependence of the absorption efficiency of the bimetallic core-shell nanotube for $R_1 = 50$ nm, $R_2 = 75$ nm and $R_3 = 125$ nm with $n_2 = 1.33$ ($\varepsilon_2 = 1.77$) and $n_4 = 1$ ($\varepsilon_4 = 1.0$). Spatial distribution of $|\mathbf{E}|^2$ at resonance wavelength with (b) $\lambda = 896.0$ nm, (c) $\lambda = 509.0$ nm, (d) $\lambda = 411.0$ nm which exhibits the dipolar (D), quadrupolar (Q) and sextupolar (S) feature, respectively.

To investigate the refractive index sensing capability, it is calculated the absorption efficiency of the core-shell nanotube with $R_1 = 50$ nm, $R_2 = 75$ nm and $R_3 = 125$ nm in vacuum ($\varepsilon_4 = 1.0$) for different value of analyte refractive indices ($n_2 = \sqrt{\varepsilon_2}$). It is found that the absorption efficiencies are spectrally redshifted as the refractive index of the analyte increased from $n_2 = 1.33$ to $n_2 = 1.45$ as shown in Fig. 3(a). This effect is related to the decreasing of LSPRs frequency for higher permittivity of the surrounding medium. One also observes the resonances wavelength depend linearly on the refractive index of the analyte as shown in Fig. 3(b). Further, using Eq. (12) one found that the refractive index sensitivity (S), given by the slope of the linear fit, for the dipole, quadrupole and sextupole modes are 664.6 nm/RIU, 300.0 nm/RIU and 179.1 nm/RIU, respectively. RIU is the refractive index unit.

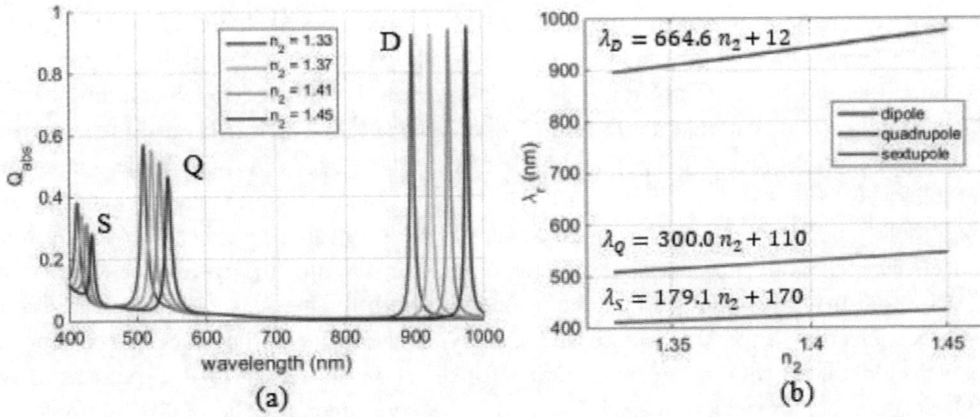

Fig. 3. (a) The spectral profile of absorption efficiency of Ag core-shell nanotube in vacuum (ε_4 = 1.0) with R_1 = 50 nm, R_2 = 75 nm and R_3 = 125 nm calculated for n_2 = 1.33–1.45. (b) Linear dependent of resonances wavelength to the analyte refractive index for the dipole (D), quadrupole (Q) and sextupole (S) modes.

3.2. Effects of the shell thickness variation

Given in Fig. 4, the effects of shell thickness on the sensing performance of the bimetallic core-shell nanotube. As can be seen in Fig. 4(a), the sensitivity (S) of refractive index sensing is relatively unchanged as the shell thickness increased. On the other hand, as seen in Fig. 4(b), the spectral linewidth of the dipole resonance varies significantly for higher shell thickness. One observes that the width of the dipole resonance decreases nonlinearly from $\Delta\lambda_D$ = 18.40 nm at shell thickness 25 nm to $\Delta\lambda_D$ = 6.26 nm at shell thickness 50 nm, while the width of quadrupole resonance decreases linearly from 17.90 nm to 10.80 nm as the shell thickness increase from 25 nm to 50 nm. In contrast to these two modes, the width of the sextupole mode increases linearly as the shell thickness increased. The significant variation of the resonances width to the shell thickness will change the figure of merit (FOM) as shown in Fig. 4 (c). The FOM of dipole mode is very sensitive to the shell thickness. It increases almost three times from 36.8 RIU^{-1} to 106.2 RIU^{-1} as the shell thickness increases from 25 nm to 50 nm. The FOM of quadrupole mode also increases almost linearly as the shell thickness increases, but with slower rate as compared to the dipole mode. Increasing the shell thickness from 25 nm to 50 nm changes the FOM of quadrupole mode from 17.4 RIU^{-1} to 27.6 RIU^{-1}. Finally, it is seen that the FOM of sextupole mode decreases slightly as the shell thickness increases.

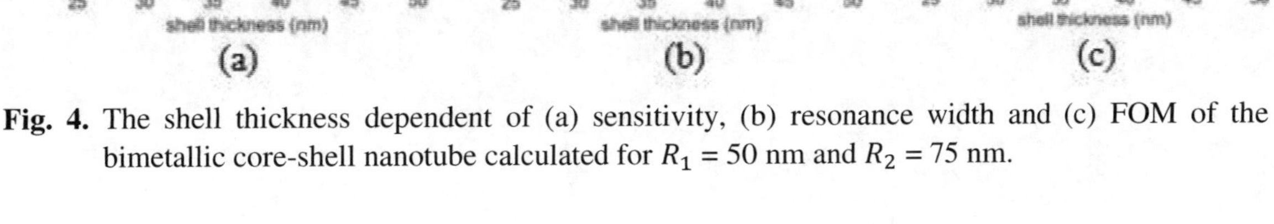

Fig. 4. The shell thickness dependent of (a) sensitivity, (b) resonance width and (c) FOM of the bimetallic core-shell nanotube calculated for R_1 = 50 nm and R_2 = 75 nm.

3.3. Effects of spacer layer thickness

The variation of sensitivity, resonance width and FOM with respect to the change of spacer layer thickness are calculated for fixed core (50 nm) and fixed shell (50 nm) and the results are given in Fig.5 below. It is found that sensitivity (S) of refractive index sensing is relatively unchanged by increasing the spacer layer thickness as demonstrated in Fig. 5 (a). Increasing the spacer layer thickness produces different effects for dipole quadrupole and sextupole mode. In the case of dipole mode, increasing the spacer layer thickness from 25 nm to 50 nm will reduce the resonance width from 6.26 nm to 5.30 nm. For the quadrupole mode, increasing the spacer layer thickness until from 25 nm to 40 nm tends to increase the resonance width, but for spacer layer more than 40 nm, the resonance width remains unchanged. Different from the other modes, the resonance width for the sextupole mode are increased drastically as the spacer layer thickness increased. Finally, as the result of the spacer layer dependence of resonant wavelength, the FOMs are also varied for different spacer layer thickness as shown in Fig. 5 (c). One observes, increasing the spacer layer thickness will increase the FOMs of dipole modes, but decrease the FOMs of quadrupole and sextupole mode.

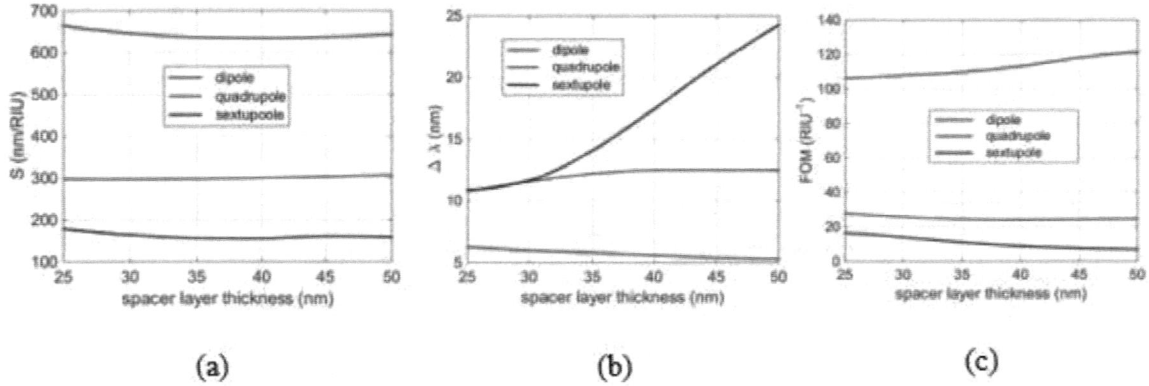

Fig. 5. Effects of spacer layer thickness on (a) sensitivity, (b) resonance width and (c) FOM of the bimetallic core-shell nanotube in vacuum.

3.4. Effects of radius variation

As shown in Fig. 6(a), increasing the core radius will increase the sensitivity monotonously for all modes. For the dipole mode, increasing the spacer layer thickness up to certain limits tends to decrease the resonance width, but after the certain value of core radius, the resonance width increases drastically as demonstrated in red line in Fig 6(b). Since the FOM depends on the resonance width, one found that as the core radius increases from 50 nm to 65 nm, the FOM increases from 106.2 to its maximum value 142.2 and then decreases significantly for the higher core radius as shown in Fig. 6(c). The resonance width of quadrupole and sexupole mode oscillate as core radius increased as shown in Fig. 6(b). Finally, from Fig. 6(c), the FOM of quadrupole and sextupole mode increased monotonously as the core radius increased.

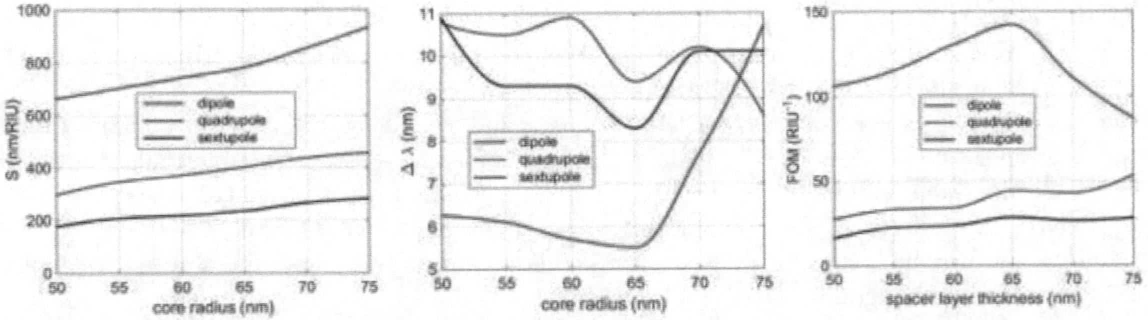

Fig. 6. Effect of core radius on (a) Senstivity, (b) resonance width and (c) FOM of the bimetallic core-shell nanotube in vacuum.

3.5. Effects of background type on the sensing performance

The background medium also affects the sensing performance. It is tabulated in Table 1, the sensitivity, resonance width and FOM of the bimetallic core-shell nanotube in vacuum and in SiO_2 background. One observes that sensing performance in SiO_2 is lower than in vacuum.

Table 1. Sensing performance of Ag core-shell nanotube in different background

Mode	in vacuum			in SiO2		
	Sensitivity	Width	FOM	Sensitivity	Width	FOM
Dipole	664.63	6.26	106.17	665.55	8.5	78.3
Quadrupole	298.6	10.8	27.65	303.63	15.9	19.1
Sextupole	178.91	11.3	15.83	177.91	27.5	6.5

4. Conclusions

We have analysed the refractive index sensing performance of analyte inside the nanotube. It is shown that the absorption spectra exhibit three peaks which has potential for multiband sensing applications. The ultrahigh sensing performance with FOM > 100 could be achieved. Further, it is found that the dipole mode has a higher sensing performance than the other modes. Finally, the sensing performance is very sensitive to the geometrical parameter and the background medium of the nanotube.

Acknowledgments

The author thanks M.O. Tjia and A.A. Iskandar for the valuable discussion in the early stage of this work. This work was supported by Penelitian DIPA FMIPA Universitas Tanjungpura Pontianak (Contract No: 2065/UN22.8/PP/2020).

References

[1] Maier SA. Plasmonics: Fundamentals and applications. Plasmonics: Fundamentals and Applications. Springer US; 2007. 1–223 p.

[2] Chen Y, Ming H. Review of surface plasmon resonance and localized surface plasmon resonance sensor, Photonic Sensors. 2012;2(1):37–49.

[3] Tam F, Moran C, Halas N. Geometrical parameters controlling sensitivity of nanoshell plasmon resonances to changes in dielectric environment. J Phys Chem B. 2004;108(45):17290–4.

[4] Zhu J, Li JJ, Zhao JW. Improve the refractive index sensitivity of coaxial-cable type gold nanostructure: The effect of dielectric polarization from the separate layer. J

Journal of Physics: Conference Series **2274** (2022) 012012 doi:10.1088/1742-6596/2274/1/012012

Nanoparticle Res 2013 15(6):1–9
[5] Azwar, A., Data, G. W. P., Soehianie, A., and Iskandar, A. A. Study of Fano feature and its size dependent variations in TE - Polarized light scattering by a silver nanorod. AIP Conf. Proc. 2015;1677:0700191-4
[6] Azwar, A. , Soehianie, A. , Iskandar, A. A. , and Tjia, M.-O. , Fano-like spectral profile in TE wave scattering by nanowire of dissipative and dispersive materials, Journal of Nonlinear Optical Physics and Materials, 2016;25(1):1600051-11.
[7] Isro, 2017 Isro, S. D. , Iskandar, A. A. , and Tjia, M. O. , Complementary high performance sensing of gases and liquids using silver nanotube, Journal of Optics (United Kingdom). 2017;19(11):11503
[8] Hendren WR, Murphy A, Evans P, O'Connor D, Wurtz GA, Zayats A V., et al. Fabrication and optical properties of gold nanotube arrays. J Phys Condens Matter 2008 S20(36):362203.
[9] Raza, S. , Toscano, G. , Jauho, A. P. , Mortensen, N. A. , and Wubs, M. , Refractive-Index Sensing with Ultrathin Plasmonic Nanotubes, Plasmonics. 2013;8(2):193–199.
[10] P. B. Johnson and R. W. Christy. Optical Constant of the Nobel Metals. Phys Rev B. 1972;6(12):4370–9.
[11] Frezza F, Mangini F, Tedeschi N. Introduction to electromagnetic scattering, part II: tutorial. J Opt Soc Am A. 2020;37(8):1300.
[12] Suryadharma, R. N. S. , Iskandar, A. A. , and Tjia, M. O. , Photonic states mixing beyond the plasmon hybridization model, Journal of Applied Physics. 2016;120(8):0431051-9.

Journal of Physics: Conference Series **2274** (2022) 012013 doi:10.1088/1742-6596/2274/1/012013

The influence of surface roughness to particle optical characteristics by a study of Chebyshev particles

Hilman Fikry and Alexander A. Iskandar*

Physics of Magnetism and Photonics Research Division, Faculty of Mathematics and Natural Sciences, Institut Teknologi Bandung, Indonesia.

*Corresponding author: iskandar@fi.itb.ac.id

Abstract. In this work, Chebyshev particles are used to study the influence of surface roughness and nonsphericity to the optical characteristics of particles by exploiting the parameters of Chebyshev particles, namely the waviness parameter and the deformation parameter. This is analyzed by comparing the scattering, extinction, and absorption cross sections, asymmetry parameter, single-scattering albedo, phase function, and linear polarization degree of different types of particles, including spheres and several types of Chebyshev particles. To calculate the optical characteristics, Extended Boundary Condition Method is utilized. The results show that there are some deviations affected by nonsphericity or surface roughness of particles, especially in relatively large size parameter, real part of refractive index is larger than 1, and imaginary part of refractive index is equal to 0. By approximating the surface roughness of particles by Chebyshev particles, larger waviness and deformation parameters tend to give larger difference from ideal particles in optical characteristics, so the optical characteristics could be controlled by modifying these parameters.

1. Introduction

A study of nonsphericity and surface roughness is important. The reason is that there are basically a lot of particles not having ideal shape but in practical calculation the shape of these particles is often assumed to be ideal (e.g., spheres or spheroids) without considering the influence of nonsphericity and surface roughness to the optical characteristics. Moreover, according to van de Hulst [1], analytical calculations using spherical shape are used in 95% of all applications. Therefore, it is important to review if a priori assumptions about ideal shape particles contribute to significant differences from the real particles which, in this case, are approximated by Chebyshev particles.

In this study, Chebyshev particles are used as the main scattering object. The shape of Chebyshev particle is a sphere deformed by means of Chebyshev polynomial order [2]. Chebyshev particles $T_n(\xi)$ have two parameters: the waviness parameter or also known as the Chebyshev polynomial order (n) and the deformation parameter (ξ) [3]. Chebyshev particles are useful for the study of particle's surface roughness as it could be approximated by Chebyshev particles with small deformation parameter [4,5]. Another benefit is that Chebyshev particles could represent nonspherical particles, therefore they could be useful to study the influence of particle concavity to the scattering characteristics of particles [6]. Although there are another particle shapes that could be used to approximate the surface roughness and the concavity of particles, such as particles with random Gaussian surface [7], Chebyshev particles have azimuthal symmetry so that the calculations are more efficient and accurate in T-matrix based on Extended Boundary Condition Method (EBCM). Thus, Chebyshev particles can be used in many fields, ranging from nanotechnology [8–10], astronomy [11–13], aerosol [14–18], biology [19,20], and climatology [21].

The next sections are organized as follows. First, we elaborated the theory of T-matrix (EBCM) as the theoretical method to calculate the optical characteristics of Chebyshev particles. Then, we presented and discussed the results of our analysis about comparisons in physical and geometrical parameters of the particles. Finally, we concluded several remarks about our findings.

2. Extended Boundary Condition Method

In this method, incident plane waves are assumed. The idea is to expand incident and scattered fields into vector spherical wave functions:

$$\mathbf{E}^{\mathrm{inc}}(\mathbf{r}) = \sum_{n=1}^{\infty}\sum_{m=-n}^{n}[a_{mn}\mathrm{Rg}\mathbf{M}_{mn}(k_1\mathbf{r}) + b_{mn}\mathrm{Rg}\mathbf{N}_{mn}(k_1\mathbf{r})], \tag{1}$$

$$\mathbf{E}^{\mathrm{sca}}(\mathbf{r}) = \sum_{n=1}^{\infty}\sum_{m=-n}^{n}[p_{mn}\mathbf{M}_{mn}(k_1\mathbf{r}) + q_{mn}\mathbf{N}_{mn}(k_1\mathbf{r})], \quad r > r_{>}, \tag{2}$$

where $\mathrm{Rg}\mathbf{M}_{mn}$, $\mathrm{Rg}\mathbf{N}_{mn}$, $\mathbf{M}_{mn}$, and $\mathbf{N}_{mn}$ are vector spherical wave functions [22] and $r_{>}$ is the smallest spherical radius covering the scattering object placed in the center of laboratory reference frame. The expansion coefficients in equations (1) and (2) are related by T-matrix [23]:

$$p_{mn} = \sum_{n'=1}^{\infty}\sum_{m'=-n'}^{n'}(T^{11}_{mnm'n'}a_{m'n'} + T^{12}_{mnm'n'}b_{m'n'}), \tag{3}$$

$$q_{mn} = \sum_{n'=1}^{\infty}\sum_{m'=-n'}^{n'}(T^{21}_{mnm'n'}a_{m'n'} + T^{22}_{mnm'n'}b_{m'n'}). \tag{4}$$

By making use of vector wave spherical wave functions in surface integral equations of incident and scattered fields, the expansion coefficients in equations (1) and (2) can be written explicitly as

$$\begin{matrix} a_{mn} \\ b_{mn} \end{matrix} = k_1(-1)^m \int_{S'} \mathrm{d}S' \left\{\omega\mu_0[\hat{\mathbf{n}} \times \mathbf{H}_+(\mathbf{r}')] \cdot \begin{matrix} \mathbf{M}_{-mn}(k_1r',\vartheta,\varphi) \\ \mathbf{N}_{-mn}(k_1r',\vartheta,\varphi) \end{matrix} - ik_1[\hat{\mathbf{n}} \times \mathbf{E}_+(\mathbf{r}')] \cdot \begin{matrix} \mathbf{N}_{-mn}(k_1r',\vartheta,\varphi) \\ \mathbf{M}_{-mn}(k_1r',\vartheta,\varphi) \end{matrix}\right\}, \tag{5}$$

$$\begin{matrix} p_{mn} \\ q_{mn} \end{matrix} = -k_1(-1)^m \int_{S'} \mathrm{d}S' \left\{\omega\mu_0[\hat{\mathbf{n}} \times \mathbf{H}_+(\mathbf{r}')] \cdot \begin{matrix} \mathrm{Rg}\mathbf{M}_{-mn}(k_1r',\vartheta,\varphi) \\ \mathrm{Rg}\mathbf{N}_{-mn}(k_1r',\vartheta,\varphi) \end{matrix} - ik_1[\hat{\mathbf{n}} \times \mathbf{E}_+(\mathbf{r}')] \cdot \begin{matrix} \mathrm{Rg}\mathbf{N}_{-mn}(k_1r',\vartheta,\varphi) \\ \mathrm{Rg}\mathbf{M}_{-mn}(k_1r',\vartheta,\varphi) \end{matrix}\right\}. \tag{6}$$

Based on equation (5) and (6), T-matrix is obtained as

$$\mathbf{T}(P) = -(\mathrm{Rg}\mathbf{Q})\mathbf{Q}^{-1}, \tag{7}$$

where explicit forms of matrix Rg**Q** and **Q** are found in [3].

With the T-matrix, we calculated several optical characteristics of particles. First, the scattering, absorption, and scattering cross sections, which provide information about the scattered power, absorbed power, and power of both effects consecutively, are calculated as follows:

$$\langle C_{\mathrm{sca}}\rangle = \frac{2\pi}{k_1^2}\sum_{n=1}^{\infty}\sum_{m=-n}^{n}\sum_{n'=1}^{\infty}\sum_{m'=-n'}^{n'}\sum_{k=1}^{2}\sum_{l=1}^{2}\left|T^{kl}_{mnm'n'}(P)\right|^2, \tag{8}$$

$$\langle C_{\mathrm{ext}}\rangle = -\frac{2\pi}{k_1^2}\mathrm{Re}\sum_{n=1}^{\infty}\sum_{m=-n}^{n}[T^{11}_{mnmn}(P) + T^{22}_{mnmn}(P)], \tag{9}$$

$$\langle C_{\mathrm{abs}} \rangle = \langle C_{\mathrm{ext}} \rangle - \langle C_{\mathrm{sca}} \rangle. \tag{10}$$

We also calculated the single-scattering albedo ϖ, the asymmetry parameter $\langle \cos \Theta \rangle$, and the phase function a_1, which are defined as

$$\varpi = \frac{\langle C_{\mathrm{sca}} \rangle}{\langle C_{\mathrm{ext}} \rangle}, \tag{11}$$

$$\langle \cos \Theta \rangle = \frac{1}{C_{\mathrm{sca}}} \int_{4\pi} \mathrm{d}\hat{\mathbf{r}} \frac{\mathrm{d}\langle C_{\mathrm{sca}} \rangle}{\mathrm{d}\Omega} \hat{\mathbf{r}} \cdot \hat{\mathbf{n}}^{\mathrm{inc}}, \tag{12}$$

$$a_1 = \frac{4\pi}{\langle C_{\mathrm{sca}} \rangle} \frac{\mathrm{d}\langle C_{\mathrm{sca}} \rangle}{\mathrm{d}\Omega}. \tag{13}$$

The single-scattering albedo provides information about the probability that light will be scattered rather than absorbed. The asymmetry parameter is interpreted as the tendency of scattered light direction. The phase function describes the angular distribution of scattered light since it is related to the differential scattering cross section $\frac{\mathrm{d}\langle C_{\mathrm{sca}} \rangle}{\mathrm{d}\Omega}$.

We developed the EBCM program based on [24] to perform efficient and accurate calculations for Chebyshev particles. The explicit formula of Chebyshev particle shape is given by [25]:

$$r(\vartheta) = r_0[1 + \xi T_n(\cos \vartheta)], \qquad |\xi| < 1, \tag{14}$$

where r_0 is the equal-volume-sphere radius, ξ is the deformation parameter, $T_n(\cos \vartheta) = \cos n\vartheta$ is the Chebyshev polynomial of order n, and n is also known as waviness parameter.

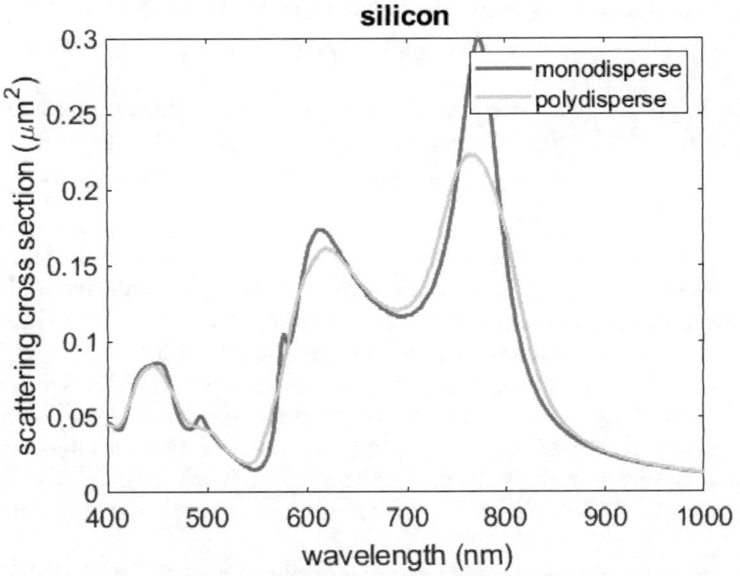

Figure 1. A comparison between monodisperse and polydisperse particle scattering cross sections. Polydisperse particle size is based on power law distribution with the effective radius 100 nm and the effective variance 0.001.

It is also important to understand the influence of size-averaging in polydisperse particles by comparing the optical characteristics of single particle or a group of single-sized particles (monodisperse) and those with different size (polydisperse). The analysis was made for the scattering cross section by using spherical dispersive silicon particles with radius 100 nm and relative refractive index based on [26]. The comparison is shown in **Figure 1**. The scattering cross section of polydisperse particles tend to smooth the curve perturbations of monodisperse particles coming from minor multipole modes. The reason is that the scattering patterns of monodisperse particles suffer from interference structure [27]. Therefore, the advantage of using polydisperse particles rather than monodisperse ones is that more meaningful comparisons could be made in optical characteristics of different particles. Besides, a group of particles is rarely found in a single size, it is found in different sizes with certain distributions instead, either in nature or in industry [28]. Nevertheless, these benefits of polydisperse particles took a cost in computational time and complexity so either monodisperse or polydisperse particles are used in this study.

3. The Analysis of Chebyshev Particle's Optical Characteristics

Our objective is to obtain the influence of Chebyshev particle parameters, namely the waviness parameter and the deformation parameter, to the optical characteristics. This can be done by comparing the scattering, extinction, and absorption cross sections, asymmetry parameter, single-scattering albedo, phase function, and linear polarization degree of different types of particles, including spheres and several types of Chebyshev particles.

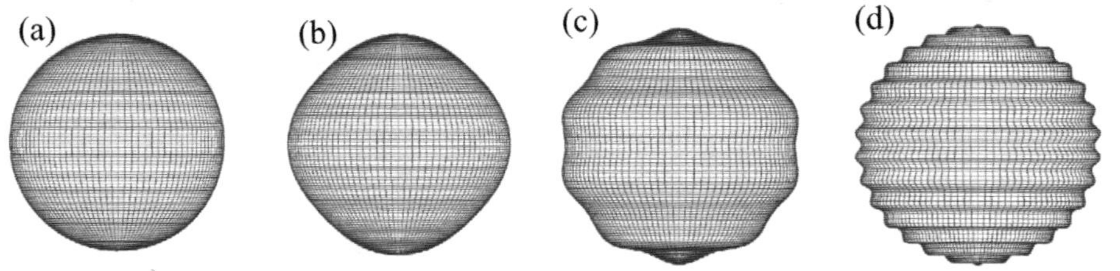

Figure 2. Particles with different waviness parameter as the scattering objects: (a) spheres, Chebyshev particles (b) $T_4(0.03)$, (c) $T_{10}(0.03)$, and (d) $T_{30}(0.03)$.

To observe the influence of waviness parameter, we use four types of particles as the scattering objects, which are illustrated in **Figure 2**. **Figure 3** shows the results of the absorption cross section and the single-scattering albedo in various size parameter, which is defined as $\frac{2\pi}{\lambda_1}r_0$, where λ_1 is the wavelength of incident light in the surrounding medium and r_0 is the equal-volume-sphere radius. In this calculation, monodisperse particles with silicon refractive index, $4.00 + 0.03i$, are used. While no meaningful difference is seen in absorption cross section and single-scattering albedo of Chebyshev particles $T_4(0.03)$, the influence of waviness parameter is observed in $T_{10}(0.03)$ and $T_{30}(0.03)$ with relatively large size parameter. The incremental differences from $T_4(0.03)$ to $T_{30}(0.03)$ prove that they are systematic deviations. The deviation in absorption cross section is observed considering that the surface area increases when the waviness and size parameters increase. By the analogy of blackbody, this increase, in turn, will increase the light absorption because there is more light interacting with the object. The increase of absorption, then, decreases the single-scattering albedo through equation (10) and (11) as the influence of waviness parameter is more apparent in the absorption rather than scattering phenomenon.

To get into practical conditions, we used dispersive silicon particles with the effective radius 500 nm and the effective variance 0.01 in polydisperse group to calculate the extinction efficiency factor and the asymmetry parameter so that the influence of waviness parameter in various wavelength could be observed clearly, as shown in **Figure 4**. We found that there is an increase of extinction efficiency factor as waviness parameter increase in smaller wavelength. This comes from the combined effect of scattering and absorption, which both show positive correlation to the waviness parameter. A deviation in smaller wavelength is expected as it simply means larger size parameter. We also found that a significant deviation is observed in asymmetry parameter of $T_{30}(0.03)$, although the other types of particles seem to be indistinguishable. This finding can be explained by **Figure 5**(a), which shows the phase function with respect to scattering angle. Different patterns are observed in the phase function of different particle types. While maintaining its pattern width, phase function pattern of $T_{30}(0.03)$ seem to have longer length than its counterparts. This indicates that more light is scattered exact-forwardly rather than arbitrarily which, in turn, increase the asymmetry parameter of $T_{30}(0.03)$. This interpretation is in accordance with equation (12) and (13). Different linear polarization degree patterns of different particle types are also observed from **Figure 5**(b). The linear polarization degree of $T_{30}(0.03)$ larger than its counterparts in most of the scattering angle means that $T_{30}(0.03)$ particles have more capability to create linear polarization from unpolarized light.

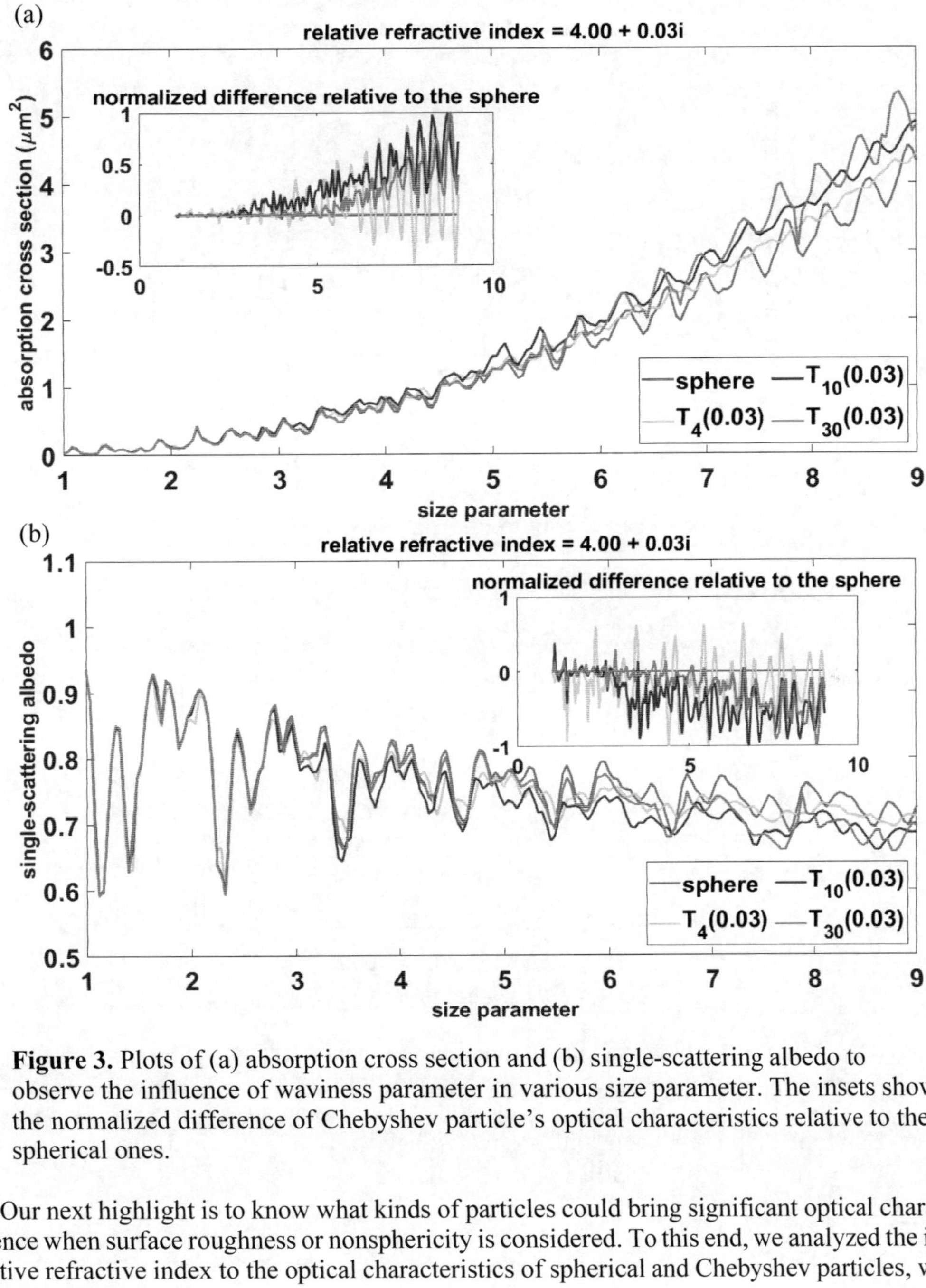

Figure 3. Plots of (a) absorption cross section and (b) single-scattering albedo to observe the influence of waviness parameter in various size parameter. The insets show the normalized difference of Chebyshev particle's optical characteristics relative to the spherical ones.

Our next highlight is to know what kinds of particles could bring significant optical characteristic difference when surface roughness or nonsphericity is considered. To this end, we analyzed the influence of relative refractive index to the optical characteristics of spherical and Chebyshev particles, which are illustrated in **Figure 6.**

Journal of Physics: Conference Series **2274** (2022) 012013 doi:10.1088/1742-6596/2274/1/012013

Figure 7 shows the result of $T_7(0.03)$ particles' scattering efficiency factor deviation from spherical particles. Significant deviation is observed when real part of refractive index is larger than 1 and imaginary part of refractive index is equal to 0. It is obvious that the real part of refractive index should be larger than 1 to see the influence of surface roughness or nonsphericity. As for the imaginary part of refractive index, the optical characteristics of very absorbing objects with various shapes tend to be indistinguishable considering that most of the light will be absorbed rather than scattered.

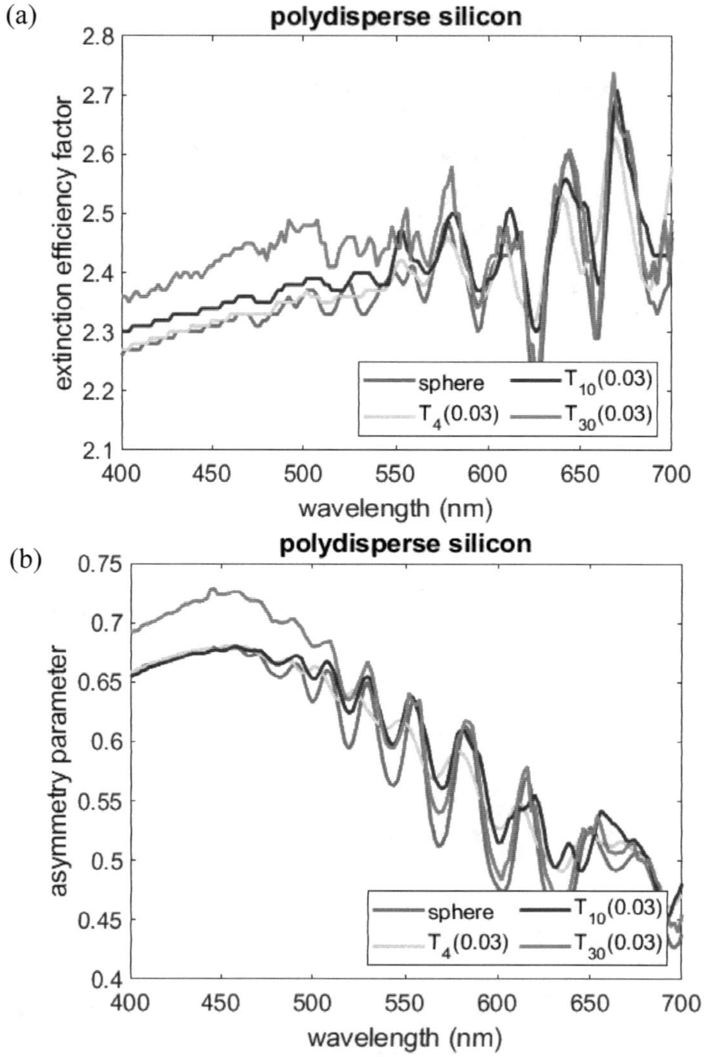

Figure 4. Plots of (a) extinction efficiency factor and (b) asymmetry parameter to observe the influence of waviness parameter in various wavelength. The scattering object is polydisperse silicon with the effective radius 500 nm and the effective variance 0.01.

Journal of Physics: Conference Series **2274** (2022) 012013 doi:10.1088/1742-6596/2274/1/012013

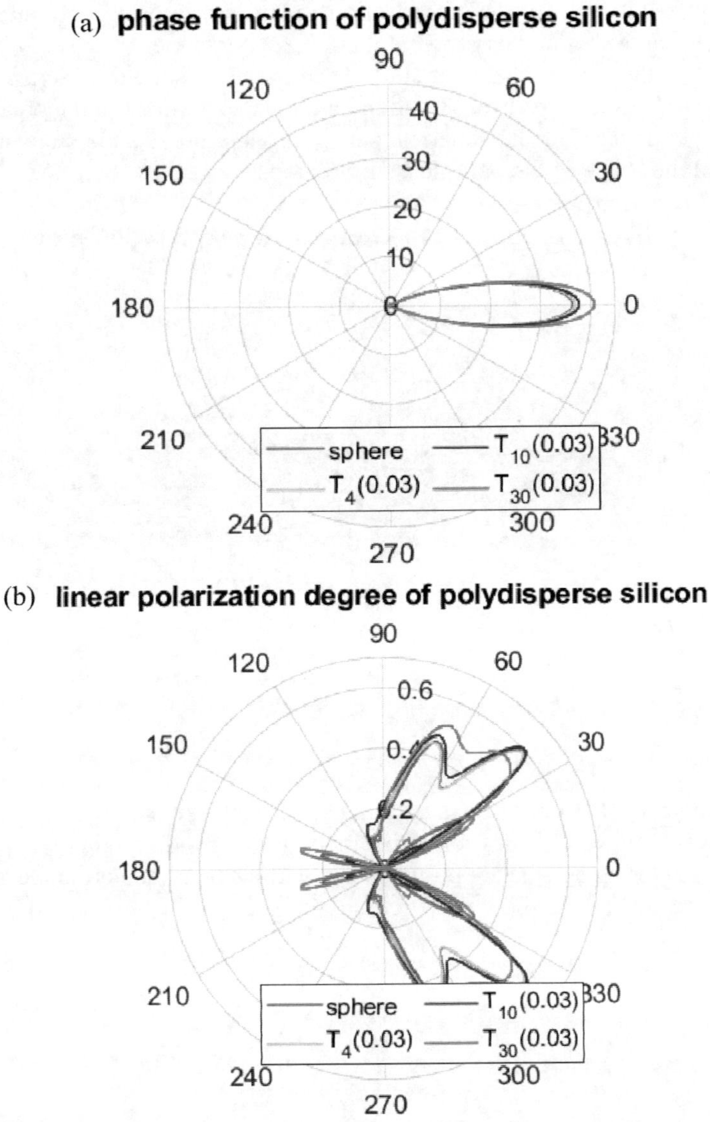

Figure 5. Plots of (a) phase function and (b) linear polarization degree with respect to scattering angle. The scattering object is polydisperse silicon with the effective radius 500 nm and the effective variance 0.01 at the wavelength 500 nm.

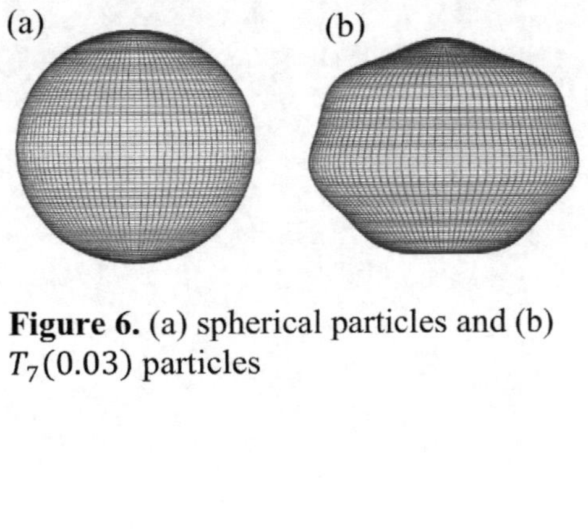

Figure 6. (a) spherical particles and (b) $T_7(0.03)$ particles

Finally, we calculated the absorption cross section and the asymmetry parameter of T_7 Chebyshev particles with different deformation parameters and size parameters to study the influence of deformation parameters to the optical characteristics. The results are shown in **Figure 8**. We found that the influence of deformation parameter is observed in larger size parameter. An increase of deformation parameter results in larger deviation from spherical particles. This can be understood as the deformation create concavity so that the shape of the particles is more nonspherical.

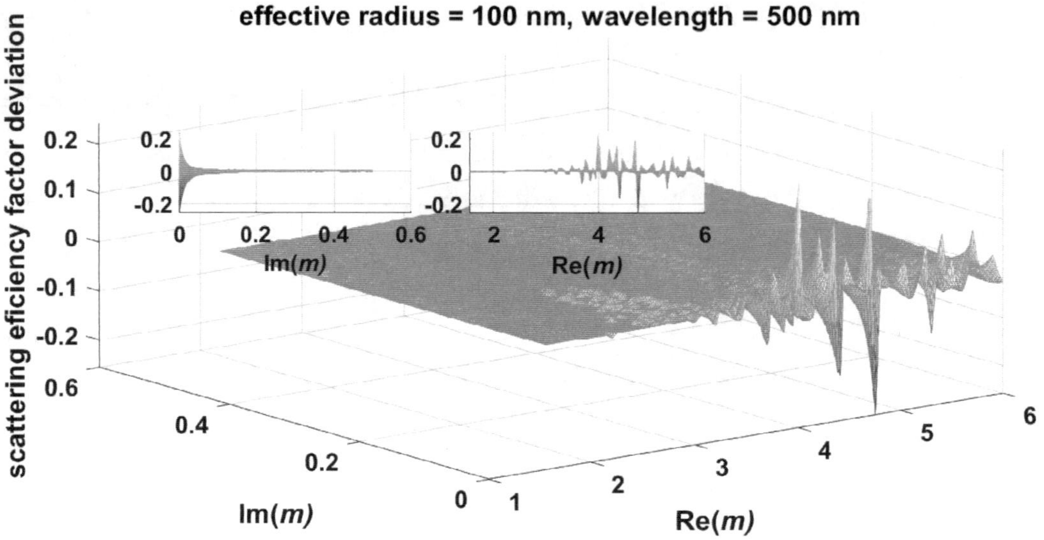

Figure 7. A plot of $T_7(0.03)$ particles' scattering efficiency factor deviation from spherical particles with respect to the real part of refractive index $\mathrm{Re}(m)$ and imaginary part of refractive index $\mathrm{Im}(m)$. The scattering objects are polydisperse particles with the effective radius 100 nm and the effective variance 0.01 at the wavelength 500 nm.

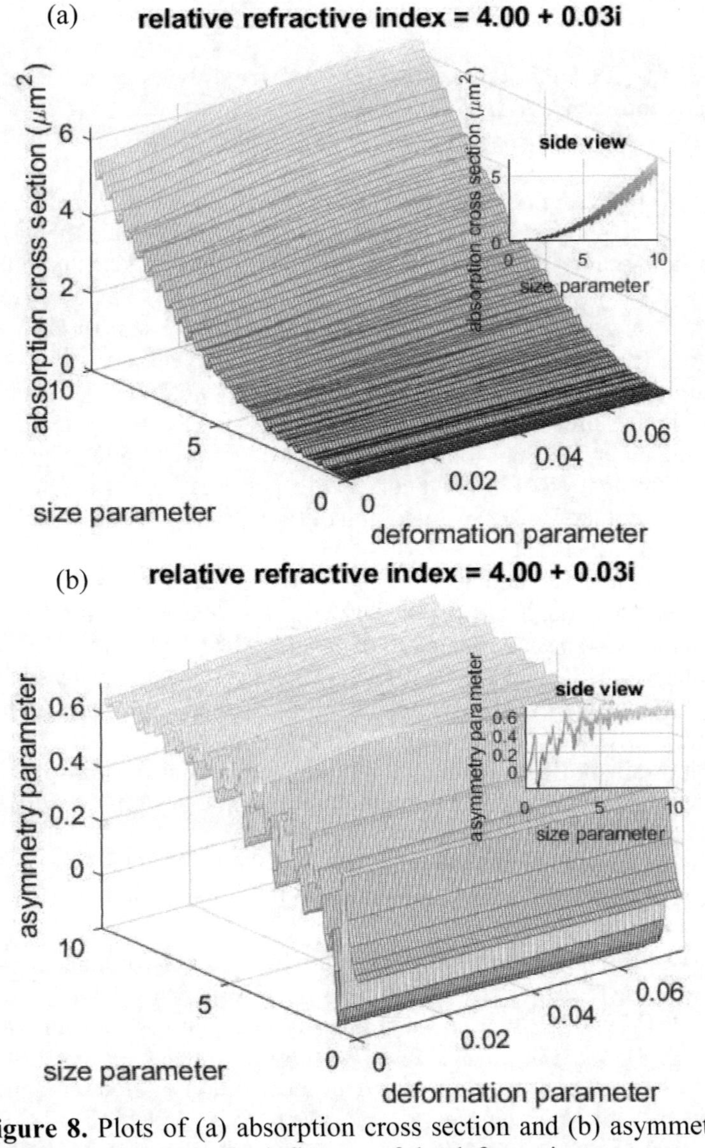

Figure 8. Plots of (a) absorption cross section and (b) asymmetry parameter to observe the influence of the deformation parameter in various size parameter. The scattering objects are monodisperse particles with silicon refractive index, 4.00 + 0.03i.

4. Conclusions

We have already seen that there are some deviations affected by nonsphericity or surface roughness of particles, especially in relatively large size parameter, real part of refractive index is larger than 1, and imaginary part of refractive index is equal to 0, so one should take a careful assumption when calculating particle's optical characteristics with ideal-shaped particles. The deviations appear to be in many optical characteristics: scattering, extinction, and absorption cross sections, asymmetry parameter, single-scattering albedo, phase function, and linear polarization degree. By approximating the surface roughness of particles by Chebyshev particles, larger waviness and deformation parameters tend to give larger difference from ideal particles in optical characteristics, so the optical characteristics could be controlled by modifying these parameters. Nevertheless, Chebyshev particles could not perfectly approximate the surface roughness of particles due to their nature of regular shape patterns, so a study of another particle shape is needed to approximate the surface roughness better.

Acknowledgments
This work is supported by PDUPT Research Grant 2021 from the National Research and Innovation Agency of Indonesia (contract no. 2/E1/KP.PTNBH/2021).

References

[1] van de Hulst H C 1981 *Light Scattering by Small Particles* (New York: Dover Publications)
[2] Mugnai A and Wiscombe W J 1986 Scattering from nonspherical Chebyshev particles I: cross sections, single-scattering albedo, asymmetry factor, and backscattered fraction *Appl. Opt.* **25** 1235
[3] Mishchenko M I, Travis L D and Lacis A a 2002 *Scattering, Absorption, and Emission of Light by Small Particles* (Cambridge: Cambridge University Press)
[4] Rother T, Schmidt K, Wauer J, Shcherbakov V and Gayet J F 2006 Light scattering on Chebyshev particles of higher order *Appl. Opt.* **45** 6030–7
[5] Kahnert M and Rother T 2011 Modeling optical properties of particles with small-scale surface roughness: combination of group theory with a perturbation approach *Opt. Express* **19** 11138
[6] Mugnai A and Wiscombe W J 1980 Scattering of Radiation by Moderately Nonspherical Particles *J. Atmos. Sci.* **37** 1291–1307
[7] Kahnert M, Nousiainen T, Thomas M A and Tyynelä J 2012 Light scattering by particles with small-scale surface roughness: Comparison of four classes of model geometries *J. Quant. Spectrosc. Radiat. Transf.* **113** 2356–67
[8] Muñoz O, Volten H, Hovenier J W, Min M, Shkuratov Y G and Jalava J P 2006 Experimental and computational study of light scattering by irregular particles with extreme refractive indices : hematite and rutile *A&A* **446** 525–35
[9] Kahnert M, Nousiainen T and Mauno P 2011 On the impact of non-sphericity and small-scale surface roughness on the optical properties of hematite aerosols *J. Quant. Spectrosc. Radiat. Transf.* **112** 1815–24
[10] Aslan M M, Mengüç M P, Manickavasagam S and Saltiel C 2006 Size and shape prediction of colloidal metal oxide MgBaFeO particles from light scattering measurements *J. Nanoparticle Res.* **8** 981–94
[11] Mishchenko M I 1989 Interstellar light absorption by randomly oriented nonspherical dust grains *Sov. Astron. Lett.* **15** 299–302
[12] Mishchenko M I 1990 Extinction of Light by Randomly-Oriented Non-Spherical Grains *Astrophys. Space Sci.* **164** 1–13
[13] Mishchenko M I and Dlugach J M 2008 Weak localization of electromagnetic waves and radar polarimetry of Saturn's rings *Mon. Not. R. Astron. Soc.* **389** 1665–74
[14] Auger J-C, Fernandes G E, Pan Y-L, Aptowicz K and Chang R K 2007 Influence of Micro-particle Surface Roughness on TAOS Patterns: Experimental and Theoretical Studies *PIERS Online* **3** 897–9
[15] Auger J C, Fernandes G E, Aptowicz K B, Pan Y L and Chang R K 2010 Influence of surface roughness on the elastic-light scattering patterns of micron-sized aerosol particles *Appl. Phys. B Lasers Opt.* **99** 229–34
[16] Bankman I N, Giles J W, Chan S C and Reed R A 2004 Model of lidar range-Doppler signatures of solid rocket fuel plumes *Laser Radar Technol. Appl. IX* **5412** 208
[17] Reed R A, Nolen M K, Zolensky M E, Bankman I, Giles J, Lang P, Mishchenko M I and Ross M N 2005 Lidar backscatter properties of Al2O3 rocket exhaust particles *J. Spacecr. Rockets* **42** 711–5
[18] Chylek P, Ramaswamy V, Cheng R and Pinnick R G 1981 Optical properties and mass concentration of carbonaceous smokes *Appl. Opt.* **20** 2980
[19] Lazarevich A K, Thomas M E, Duncan D D and Mitchell C A 2004 Determination of bacterial aerosol spectral cross sections *Chem. Biol. Sens. V* **5416** 194

[20] Pruvost J, Van Vooren G, Le Gouic B, Couzinet-Mossion A and Legrand J 2011 Systematic investigation of biomass and lipid productivity by microalgae in photobioreactors for biodiesel application *Bioresour. Technol.* **102** 150–8
[21] Baran A J, Gayet J F and Shcherbakov V 2012 On the interpretation of an unusual in-situ measured ice crystal scattering phase function *Atmos. Chem. Phys.* **12** 9355–64
[22] Stratton J A 2007 *Electromagnetic Theory* (Hoboken: John Wiley & Sons)
[23] Waterman P C 1971 Symmetry, unitarity, and geometry in electromagnetic scattering *Phys. Rev. D* **3** 825–39
[24] Mishchenko M I and Travis L D 1998 Capabilities and Limitations of a Current Fortran Implementation of the T-Matrix Method for Randomly Oriented, Rotationally Symmetric Scatterers *J. Quant. Spectrosc. Radiat. Transf.* **60** 309–24
[25] Wiscombe W J and Mugnai A 1986 Single Scattering From Nonspherical Chebyshev Particles: a Compendium of Calculations. *NASA Ref. Publ.*
[26] Aspnes D E and Studna A A 1983 Dielectric functions and optical parameters of Si, Ge, GaP, GaAs, GaSb, InP, InAs, and InSb from 1.5 to 6.0 eV *Phys. Rev. B* **27** 985–1009
[27] Mishchenko M I, Travis L D and Mackowski D W 1996 T-Matrix Computations of Light Scattering by Nonspherical Particles: A Review *J. Quant. Spectrosc. Radiat. Transf.* **55** 535–75
[28] Penttilä A and Lumme K 2004 The effect of particle shape on scattering - A study with a collection of axisymmetric particles and sphere clusters *J. Quant. Spectrosc. Radiat. Transf.* **89** 303–10

AUTHOR INDEX